Rural Settlements

David Barker

Advanced
Topic*Master*

Series editor
Michael Raw

For Susan, Lizzie, Al and Ed

Philip Allan Updates, an imprint of Hodder Education, part of Hachette Livre UK, Market Place, Deddington, Oxfordshire OX15 0SE

Orders
Bookpoint Ltd, 130 Milton Park, Abingdon, Oxfordshire OX14 4SB
tel: 01235 827720
fax: 01235 400454
e-mail: uk.orders@bookpoint.co.uk
Lines are open 9.00 a.m.–5.00 p.m., Monday to Saturday, with a 24-hour message answering service. You can also order through the Philip Allan Updates website: www.philipallan.co.uk

ISBN 978-1-84489-613-4

First printed 2008
Impression number 5 4 3 2 1
Year 2012 2011 2010 2009 2008

Printed in Spain

Hachette Livre UK's policy is to use papers that are natural, renewable and recyclable products and made from wood grown in sustainable forests. The logging and manufacturing processes are expected to conform to the environmental regulations of the country of origin

Contents

Introduction

This book provides a detailed analysis of rural settlement systems, the ways in which they change and the planning responses. It is written mainly for sixth-form students studying AS and A-level geography but the content may also be of value to first-year university students.

Rural settlements are highly significant. At the global scale, in terms of numbers alone, they are home to more people than urban areas. In India nearly 800 million people were estimated to live in rural settlements in 2007; the figure for China was 827 million. At the national scale, it is increasingly recognised that planning policies that sustain rural settlements and rural economies also play a major role in alleviating the problems of the large urban areas, especially in many less economically developed countries (LEDCs).

Even in countries where the proportion of rural dwellers is much lower, such as the UK, rural settlements are of great importance. One reason is that they are inextricably linked to the fundamental economic activities of rural areas, such as farming. Another is that, as central places, they provide vital services for significantly large and mobile populations living in rural regions and urban hinterlands.

In more economically developed countries (MEDCs) over the last 40 years, many villages and small market towns have experienced socioeconomic change. As a result, their economic sustainability and the management of social problems and conflicts that arise have become crucial issues. If rural settlements and settlement hierarchies are to be managed effectively, it is important to identify and understand the causes of change and the consequences, such as rural deprivation.

This *Advanced TopicMaster* covers these processes of change and related issues in three main sections. The first section (Chapters 1 and 2) deals with the characteristics of rural settlements and rural settlement patterns; the second section (Chapter 3) explains the role of rural settlements within the settlement hierarchy; the third section (Chapters 4 to 7) examines the effects of change and the planning responses, with reference to detailed case studies at different scales.

Each chapter has been designed to meet the specific needs of examination candidates. This includes definitions of geographical terms, descriptions of settlement patterns and explanations of human and physical factors that have influenced rural settlements. A range of case studies provides up-to-date factual knowledge, applied to every topic; this information is supported by a variety of photographs, maps, diagrams and tables. Current trends and issues concerning depopulation and counterurbanisation form a significant part of the book. Planning policies are assessed and their relative successes are evaluated.

The activity sections are an integral part of each chapter and include skills exercises, such as OS map interpretation and the analysis of statistical maps and diagrams. Data-response questions and essay titles are also included. There are suggestions for further research on the internet, and advice is given regarding fieldwork. The personal study of a rural settlement, settlement hierarchy or an aspect of recent change could form the basis of an effective coursework investigation or dissertation.

David Barker

1 Defining rural settlements

What is a rural settlement?

It is difficult to give a simple definition. A simplistic, traditional view is that rural settlements are relatively small in size with old thatched stone houses and most of the population employed in agriculture. This may be the case in some remote areas or in LEDCs but in the twenty-first century it cannot be said of many villages and small market towns, especially in MEDCs.

The problem of definition exists because of the great variety of rural settlements. Rural settlements are closely related to the physical environments in which they are situated and to the rural economies of which they are part. In the UK alone, where there are many different landscapes and rural economies, rural settlements display marked differences in their built environments and in their socioeconomic structures.

The difficulty is compounded by the rapid social and economic changes that have taken place in many rural areas over the last 40 years. During this period, in MEDCs for example, the effects of internal migration have led to the differences between rural and urban societies becoming much more indistinct.

Activity 1

Figure 1.1 **Six different rural settlements**

Ralegan Siddhi, India   Hualcayan, Peru

Prasan Firodia   Lizzie Barker

Activity 1 (continued)

Brandeston, Suffolk, UK

Framlingham, Suffolk, UK

Hafod-y-Llan, Gwynedd

Cleland, Barbados

Study Figure 1.1, which shows six different rural settlements. Identify the characteristics of each of the settlements shown and suggest reasons for the differences.

Each characteristic on its own is not adequate as a measure or defining characteristic of a rural settlement. The more frequently cited indices in defining rural settlements are outlined below.

Indices used in the definition of rural (and urban) settlements

Population size

Most countries have established a clear distinction between rural and urban settlements on the basis of population size. In many Latin American and west African countries, a rural settlement has 2000 people or less; the figure for Iceland and Norway is 200; in the Czech Republic it is 2000; in the USA it is 2500; and in Italy a rural settlement has a population below 10 000.

Clearly there is much global variation and the figures appear to be arbitrary. In most instances the definition is not that simple. Often it is linked to the specific problems and planning issues of a particular country.

For example, in India the definition is complex. Urban populations are defined first; the remainder are designated as rural. All populations living within a specified urban authority (such as a municipal corporation or a cantonment board) are treated as urban. In addition, an urban area must have a minimum population of 5000; 75% of the male working population must be engaged in non-agricultural employment; and the population density must be at least $400 \, km^{-2}$.

In the USA, largely because of rapid change in the rural–urban fringe, revision of the urban/rural definition has been necessary at each decennial census. Since 1950, the USA has adopted the concept of the urbanised area (UA), which has a minimum of 50000 persons. In addition, other delineated places called urban clusters (UC), which have at least 2500 persons, are 'urban' if they include contiguous densely settled territory of at least 1000 people per square mile. Therefore 'rural' consists of all territory, population and housing units outside the UAs and UCs.

Population structure

Rural settlements may have a distinctive age–sex structure, but there is no single generic type. Population structure tends to be related to the in- or out-migration that has affected the rural settlement. For example, there could be evidence of ageing in rural settlements affected by either depopulation or retirement migration. Similarly, rural settlements close to the rural–urban fringe may have a more youthful population structure as a result of counterurbanisation. Rural settlements have traditionally had a more homogeneous ethnic structure than urban areas; this has also changed in the last 30 years.

Employment structure

In urban areas, secondary and tertiary employment is more significant than primary employment. Because of the traditional economy of most rural areas, the employment structure of rural settlements has been dominated by primary activity — for example, farming, forestry and mining. While this may still be the case today in remoter rural settlements of MEDCs (and in the subsistence societies of LEDCs), the pattern is changing. In countries such as the USA and the UK, there has been a distinct shift in the employment structure of many villages within commuting distance of large urban areas. Increased personal mobility and improved public transport systems have enabled many new rural residents to travel to work, retaining their tertiary employment in the city.

Service provision

An obvious feature of rural settlements is that they tend to have fewer shops and services than urban areas; there is less demand and threshold populations cannot be met. Rural settlements are also likely to have a higher proportion of low-order shops than urban areas. A greater proportion of high-order functions, such as hospitals, and shops selling durable items, tends to be found in urban settlements. The number and type of shops and services (or a related measure, such as total retail floor space) is a useful basis for classification of rural settlements within the settlement hierarchy. However, there are anomalies. Can you explain why?

Social characteristics

The social characteristics of rural settlements are not easy to define either. The traditional village was said to have a strong community spirit compared with the more diverse populations of urban areas. Until about 1960, villagers in the UK conducted their lives in similar ways (for example, in their patterns of behaviour, such as going to church) and they had similar attitudes towards change. There would be strong family ties and limited movement. This has broken down in many instances: the village may have expanded; some newcomers may have been a source of social conflict; urban attitudes and lifestyles may have been imported. Is it possible to quantify or measure these characteristics? How useful are they in identifying rural settlements and rurality?

Land use

Definitions are just as difficult in terms of land use. Land use surrounding rural settlements tends to be mainly agricultural, extensive, and of limited variety. Around an urban area, the demands of the urban population may have led to greater diversity in the surrounding land use: recreation, new roads, modern industry, business parks and superstores.

Activity 2

Figure 1.2 **The small market town of Framlingham, Suffolk in 1953**

© English Heritage

Study Figure 1.2. Describe the characteristic features of Framlingham in 1953. In what ways might this small market town have changed as a result of population growth in the last 40 years?

Rurality

The definition of a rural area or rural region is just as problematic as that of the individual rural settlement. The term **rurality** is used to describe the degree to which an area is rural rather than urban. An index of rurality, taking into account 16 different indices, was devised by Paul Cloke (1977) ('An Index of Rurality for England and Wales', *Regional Studies*, Vol. 11, pp. 31–46). This was initially based on 1971 census data. Examples of measures used include: distance from a town with a population of over 50 000; occupational structure; retired population; population density; and commuter population. As a result, in addition to the 'urban' category, four different types of rural area were identified: 'extreme rural'; 'intermediate rural'; 'intermediate non-rural'; and 'extreme non-rural'. By mapping these areas, a rurality map of England and Wales was produced. This work was updated by Cloke and Edwards, and by Harrington and O'Donoghue, after the 1981 and 1991 censuses respectively.

This research has led to the development of a useful model — that is, the degree of rurality varies spatially in a continuum from urban area to remote countryside. The two rural extremes are described in Table 1.1. In reality, in England and Wales, this continuum is not one of regular change; there are many anomalies. Similar difficulties of definition have been experienced in the USA. The US Census Bureau confirmed this seemingly intractable problem in its definition for the 1990 census, stating: 'there is both urban and rural territory within both metropolitan and non-metropolitan areas'.

Table 1.1 Characteristics of 'extreme rural' and 'extreme non-rural'

Extreme rural: problematical remote rural areas	Extreme non-rural: extreme urban pressure
• Areas tend to be remote from large urban centres	• This type of area is close to the urban district, often located in the rural–urban fringe
• A relatively low percentage of the population commutes to work	• A high proportion of the population commutes to work, living in dormitory settlements
• There is out-migration, a low percentage of young working population and a high percentage of elderly	• The age structure includes mainly young working ages
• Land use is predominantly related to primary activity, especially agriculture	• Land use is a mixture of agriculture and a high proportion of recreational usage around the city, such as sports centres, golf courses, equestrian centres
• The area may be protected as part of a national park	• This area may include green-belt land
• The rural settlements may be largely unaltered in their physical structure	• Residents of the villages have a high degree of personal mobility and command high wages
• There is a relatively high proportion of second homes, declining service provision and rural deprivation	• Village services decline, since many residents shop in the urban area or out of town

Can you suggest any other indices that would help in the definition of rural settlements and the degree of rurality?

Tranquillity

Could tranquillity be equated with rurality? Tranquillity is assessed by the Campaign to Protect Rural England (CPRE). This organisation has devised a tranquillity score based on 44 factors. The newspaper article in Figure 1.3 outlines some of the relevant issues.

Figure
1.3
Newspaper article (*The Times*, 10 September 2007)

How the spread of towns reduced England's green and peaceful land

The English countryside is shrinking rapidly, and much of it could disappear within 80 years unless there are curbs on new developments.

Alarm over the loss of undisturbed areas of the landscape is being raised today by the Campaign to Protect Rural England (CPRE). It has commissioned new maps charting the pace of construction that has changed the landscape since the early 1960s.

The striking images show that almost 50 per cent of England is now disturbed by roads, industrial developments, out-of-town retail and business parks and new housing estates.

The "intrusion" maps show exactly how the growth in motorways and roads, power stations, airports, railway lines, power lines, wind farms, mines and quarries has affected the countryside.

According to the CRPE, only 26 per cent of England's land area had been disturbed by urban intrusion before the 1960s. This grew to 41 per cent by the early 1990s, and this year to almost 50 per cent, 25,614 square miles (66,399 sq km). The extent of incursion may be even greater, however, as the mapping exercise did not take into account the impact of aircraft noise.

The maps show that the main area of remaining, undisturbed countryside is in the National Parks — Dartmoor, Exmoor, the Yorkshire Dales, the North York Moors, the Lake District and Northumberland — which have strong planning controls. There are also large swaths of deep, unfragmented countryside along the Marches of Herefordshire and Shropshire, and in the North Pennines.

Land in the South East is under the most pressure. In the past decade, 320 square miles (830 sq km) of countryside have been affected by the impact of new developments, the equivalent of building over the area of Greater London every two years.

The CPRE is publishing the maps in the hope that they help to rein in the Government's shake-up of the planning system. Gordon Brown has signalled that a planning reform Bill will be included in the next legislative programme. The reforms would make it easier to build new infrastructure, particularly controversial new airports, and expand existing sites.

51,300
square miles, England's land mass

13,100
square miles, affected by development before the 1960s

20,820
square miles in 1990s

25,614
square miles affected by noise or visual intrusion today

Campaigners want ministers to recognise that the countryside is diminishing, and that building on only 1 or 2 per cent of existing countryside would have a significant effect on the character of the landscape. The CPRE is calling for more building on brown-field sites, and for greater promotion of public transport. Shaun Spiers, the group's chief executive, said: "The countryside is one of our greatest national assets. I am sure that the Government wants to protect it — but these maps show the current pace of development is seriously eroding our countryside.

He added: "The impact of development spreads way beyond its immediate footprint. More must be done to protect what is left from further fragmentation.

"The Government must act across the board to demonstrate that it takes the future of the countryside seriously. Unless it does so, for children alive today, much of our remaining undisturbed countryside will become a distant memory in their lifetimes."

Do the maps produced by CPRE (Figure 1.4) help to distinguish between rural and urban areas?

Figure 1.4	**Erosion of England's tranquillity**

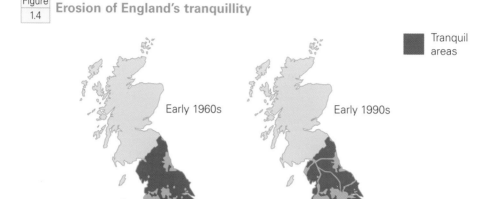

Tranquil areas

Early 1960s   Early 1990s

Activity 3

Study Figure 1.4.

(a) With reference to the two maps shown in Figure 1.4, describe the changes in the pattern of tranquil areas in England between the early 1960s and the early 1990s.

(b) Visit the website of the Campaign to Protect Rural England at: www.cpre.org.uk/home By following the sequence Campaigns, Landscape, Tranquillity, and then National and regional tranquillity maps, you can find the most recent map of your area. Describe the differences between the map of your region on this site and the map for the early 1990s in Figure 1.4.

(c) Using the newspaper article in Figure 1.3, suggest reasons for the changes described in **(b)**.

Why is it important to define rural areas?

Authorities need detailed information when preparing management plans for rural areas. These plans are essential in sustaining natural environments and local and regional economies. They are more effective if rural areas and the types of socioeconomic problems within them (often highly localised) can be identified accurately. Two examples — the Czech Republic and England and Wales — illustrate this point.

The Czech Republic

Since the Czech Republic's accession to the EU in 2004, the Czech government has been able to divert more funding to its Agriculture and Rural Development Plan. The classification adopted (Table 1.2) has been fundamental in the correct allocation of funds for rural development.

Table 1.2 The structure of settlements according to size in the Czech Republic, 2005

Size of settlement	Number of settlements	% of settlements	Population in settlements	% of population
Up to 199	1633	26.14	200 534	1.96
200–499	2012	32.20	653 740	6.40
500–999	1293	20.69	901 546	8.82
1000–1499	449	7.19	545 888	5.34
1500–1999	225	3.60	387 968	3.80
Total 'villages'	**5612**	**89.82**	**2 689 676**	**26.32 rural**
2000–4999	368	5.89	1 122 262	10.98
5000–9999	137	2.19	932 726	9.13
10 000–19 999	68	1.09	955 227	9.35
20 000–49 999	42	0.67	1 250 363	12.23
50 000–99 999	16	0.26	1 157 242	11.32
100 000+	4	0.06	942 510	9.22
Prague	1	0.02	1 170 571	11.45
Total 'towns'	**636**	**10.18**	**7 530 901**	**73.68 urban**
Total Czech Rep.	**6248**	**100**	**10 220 577**	**100**

Source: Czech Statistical Office

Table 1.2 shows that in 2005, settlements of less than 2000 people were classified as rural. Under this classification, the country has 5612 rural settlements (i.e. nearly 90% of all Czech settlements) containing 2.69 million people (26.3% of the Czech population). The size categories appear to be arbitrary, but in fact they have been established using a complex set of socioeconomic indices. This classification is important, since it has enabled grants to be allocated correctly to the rural areas specifically in need of development.

England and Wales

An even more detailed classification of rural and urban areas was devised in a government project in 2004. It was sponsored by the Department for Environment, Food and Rural Affairs (DEFRA), the Countryside Agency (CA), the Office for National Statistics (ONS) and the National Assembly for Wales (NAW). This project, coordinated by Birkbeck College, University of London, is a complicated classification based on many indices. Economic, social and demographic characteristics have been recorded for every hectare in England and Wales. In addition, every hectare has been given a 'sparsity score' depending on the density of households. Using this information, each census Output Area (2001) has been classified as urban or rural. Further details can be found at: www.statistics.gov.uk/geography/nrudp.asp

Information at this scale is important in ensuring that planning policies can be applied consistently in rural areas throughout England and Wales. Often, rural problems do not affect an entire rural area uniformly. For example, rural poverty can be hidden in small pockets. Identification of problem areas at this small scale enables specific and effective application of rural planning policies.

An example of the use of this classification at county level can be found on the Suffolk County Council website at: www.suffolkcc.gov.uk. Suffolk County Council has mapped and graphed its Output Areas according to this classification (see Figures 1.5 and 1.6). This has been of great value in the implementation of planning policies in the county's rural districts.

| Figure 1.5 | Classification of rural and urban areas in Suffolk, 2005 |

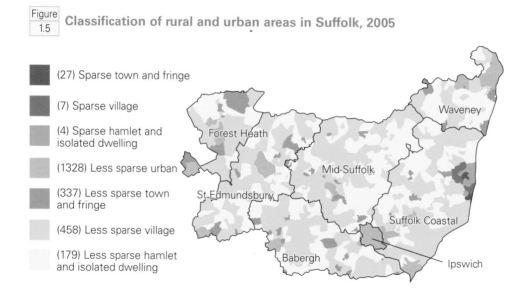

(27) Sparse town and fringe

(7) Sparse village

(4) Sparse hamlet and isolated dwelling

(1328) Less sparse urban

(337) Less sparse town and fringe

(458) Less sparse village

(179) Less sparse hamlet and isolated dwelling

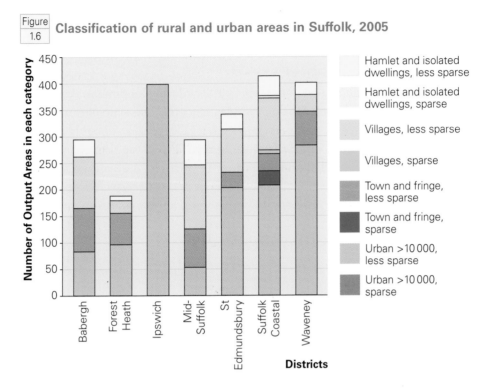

Figure 1.6 **Classification of rural and urban areas in Suffolk, 2005**

Legend:
- Hamlet and isolated dwellings, less sparse
- Hamlet and isolated dwellings, sparse
- Villages, less sparse
- Villages, sparse
- Town and fringe, less sparse
- Town and fringe, sparse
- Urban >10 000, less sparse
- Urban >10 000, sparse

Y-axis: Number of Output Areas in each category

X-axis (Districts): Babergh, Forest Heath, Ipswich, Mid-Suffolk, St Edmundsbury, Suffolk Coastal, Waveney

Key statistics are identified for each of the Output Areas and this enables detailed mapping of specific social characteristics in rural areas. Age structure, for example, is one type of data that can be related to the planning policies of Suffolk County Council. In this case, accurate identification of the spatial patterns of different age groups has helped in planning the delivery of services such as schools and public transport.

Rural settlement systems

Even though definition is a problem, use of a simple systems approach gives some commonality to the characteristics of rural settlements and the relationships between them. Rural settlements are open process–response systems — for example, they are related to the physical and economic environments in which they have evolved and they are also linked to other settlements within the hierarchy.

Inputs into the system include physical, demographic, socioeconomic and political influences; one of the most obvious is that of in-migration. These are, in effect, the processes of change, which affect the state of dynamic equilibrium within the rural settlement system. Another example is planning policies.

Outputs include the response to change. These are often evident in the specific characteristics of the settlement as it exists today, for example its built environment and its demographic structure. Other outputs might include out-migration of population to another system (perhaps urban) or physical expansion of housing estates impinging on both farming and hydrological systems. Population and services can be viewed as two examples of stores, in a state of flux, within such a system.

Many natural systems have self-adjusting mechanisms or negative feedback loops that respond to change and restore balance. In the case of the rural settlement, change such as population growth or decline is more likely to lead to further change or positive feedback unless planning responses are very successful.

Personal fieldwork investigations

Rural settlement systems can provide the basis for interesting personal investigations. Many relevant geographical themes and issues (such as migration, population structure, land use, shop and service provision, status within the hierarchy, socioeconomic characteristics and attitudes of residents) can be pursued in the context of recent change.

Possible titles for coursework with differing degrees of difficulty include:

- How and why has village V changed since 1960?
- To what extent does village W conform to the Hudson model (see Figure 7.6 on page 120) of change in a rural settlement?
- How and why do the social and economic characteristics of (three or four) rural settlements vary with distance from a large urban area?
- An evaluation of the impact of depopulation/counterurbanisation on settlement X.
- What factors have influenced the distribution of service Y in rural area Z in the last 20 years?

2 Rural settlement patterns

Rural settlement patterns are dynamic systems, evolving over long periods of time. They are the product of physical, economic and social factors, including political decisions, which influence human activities.

From the outset you should understand that **settlement pattern** refers to the spatial distribution of villages, hamlets and individual farms in a region or area. The concepts of **nucleated settlement pattern** (dominated by villages) and **dispersed settlement pattern** (dominated by hamlets and isolated farms) are explained and illustrated later in this chapter.

Site, situation and morphology

It is important to understand that each *individual* rural settlement, which is part of the overall pattern in an area, has its own particular characteristics. The distinction between the terms site and situation is fundamental in understanding the evolution of rural settlement.

- **Site** is the actual land on which an individual settlement has been built.
- **Situation** is the location of a settlement in relation to the features of the surrounding landscape.

Site

The original site of many settlements was influenced by the physical environment, the level of technology to alter that environment and the perceptions of the societies involved. In most instances, at least one of the following factors was an important consideration in the specific choice of site:

- **dry-point** — well drained and elevated above any perceived flood hazard
- **wet-point** — close to a water source for domestic and agricultural use
- **slope angle** — relatively gentle slopes were easier for the construction of dwellings
- **hill-top** — more easily defended and, in the tropics and subtropics, cooler and freer from disease (e.g. malaria) than neighbouring lowland

- **shelter** — the lee of a steep slope may have afforded shelter from prevailing winds and harsh weather conditions
- **aspect** — south-facing slopes in the northern hemisphere are favoured for agriculture and settlement, since the intensity of solar radiation and duration of sunlight are both greater than for more shaded north-facing slopes

Figure 2.1 The sites of Ufford and Eyke, Suffolk

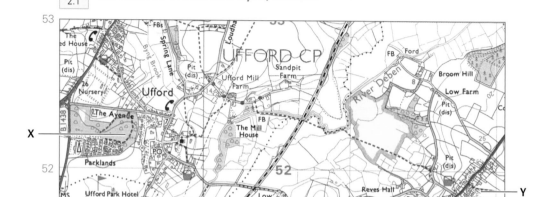

Figure 2.1 is a 1:25 000 OS map extract, which shows the site of Ufford in the valley of the River Deben. The specific evidence of the OS map shows that three factors have influenced this particular site:

- It is a dry-point site: most of the buildings are to the west of the 10 m contour, i.e. above the floodplain of the river, avoiding the area of potential flood hazard.
- It is a wet-point site: apart from the Deben itself, a regular supply of fresh water is available from a small tributary, Byng Brook (parallel to Spring Lane).
- It has gentle slopes: the 10m, 15m and 20m contours west of the church are widely spaced, indicating relatively flat land, allowing ease of building.

Activity 1

Using the contours and spot heights in Figure 2.1, draw a cross-section to show the relief of the Deben Valley between points X and Y. Annotate your diagram to show the floodplain, the river channel and the sites of Ufford and Eyke.

Factors affecting settlement site are intrinsic, whereas those affecting situation are extrinsic. Intrinsic factors are those that operate internally — they are features of the actual village site itself. Extrinsic factors are influences beyond the village site, either in the immediate vicinity or further afield.

Figure 2.2 provides a summary of the **intrinsic factors** that determine the sites of rural settlements. They are identified mainly in terms of the physical geography. The values of the particular culture, the level of technology and the perceptions of the settlers themselves are also important.

| Figure 2.2 | **Factors affecting the site of rural settlements** |

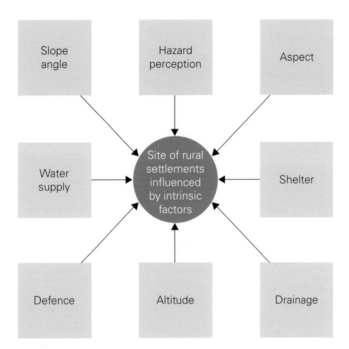

Activity 2

Conduct a simple field transect to identify the site of a particular rural settlement.
Use a large-scale OS map, such as 1:5000 (Landplan), which shows contours at 5m intervals, or 1:2500 (Superplan), which provides detail for every building.

(a) Record on the map the land use for each building/open space.

(b) Record on the map the age of building.

(c) Annotate on your map the relevant site factors for your chosen settlement.

Activity 3

The sites of Calver and Curbar

Curbar   River Derwent   Calver   (West) Calver Peak (291 m)
Calver Low (200 m)

(East)
Curbar Edge
(310 m)

D. Barker

Profile of the Derwent Valley

Figures 2.3 and 2.4 show the sites of Calver and Curbar in Derbyshire. Using the evidence of both the photograph and the valley profile:

(a) compare the sites of Calver and Curbar

(b) explain the advantages of the original site of Calver

Situation

A variety of **extrinsic factors** interact to affect the situation of rural settlements. Extrinsic factors exert an influence beyond the site itself. Figure 2.5 summarises those extrinsic factors that operate in *close proximity* to a settlement.

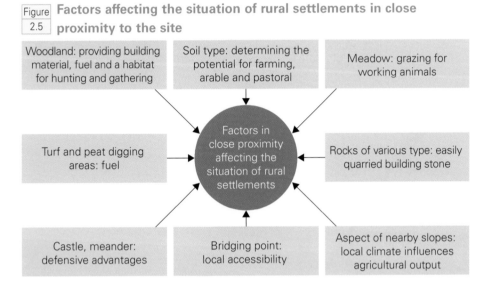

Figure 2.5 **Factors affecting the situation of rural settlements in close proximity to the site**

The situation of a settlement may also be affected by factors that apply in its *wider regional setting*. These factors include:

- nodality
- trade
- strategic position
- perceptions of the natural environment
- travelling time/distance of a village or market town from an urban area
- the juxtaposition of two distinctly different types of land use

Nodality

A nodal point is where routes converge. There may be significant differences in the degree of route connectivity of settlements in lowland plains and upland areas. Originally this may have depended on valley junctions or river confluences, and latterly on the modern road network. A good example of the influence of nodality is evident in the development of rural settlements in eastern Scotland. On the Carse of Gowrie, a lowland north of the Firth of Tay, villages are more accessible, closer together, larger in size and of greater centrality than those of the steep-sided Braes of the Carse (Sidlaw Hills). Here the settlements, little more than hamlets, are widely spaced in an area with few roads (see Figure 2.6).

Figure
2.6
The contrasting situations of settlements on the Carse of Gowrie (lowland) and on the Braes of the Carse (upland)

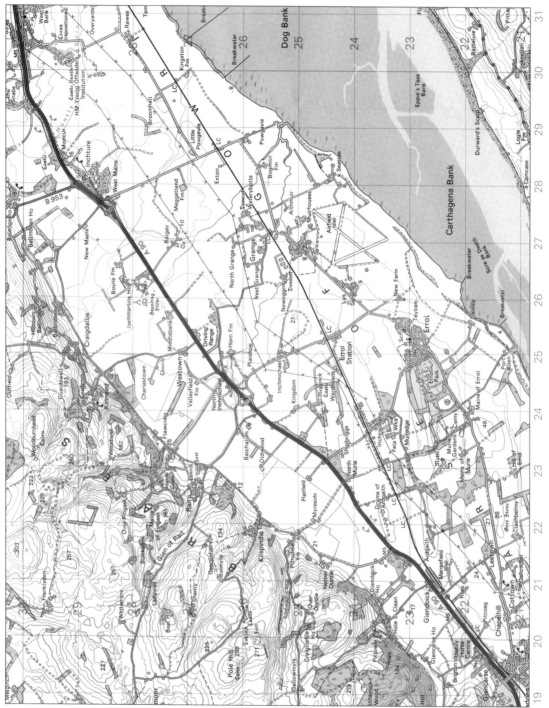

Trade

Access to a navigable river/natural or man-made harbour may help to increase the importance of a rural settlement. Porthmadog is situated on a small estuary on the south coast of the Lleyn Peninsula, on the southwest fringes of the Snowdonia National Park. Its advantageous position has enabled it to maintain its prosperity at various stages in its history as a market town, service centre, fishing port, slate port and tourist centre.

Strategic position controlling a gap in a range of hills

This may have been of military importance or for communications both along the ridgeway and through the valley. The village of Corfe Castle in Dorset is well known for its castle, which enabled control of a gap in a well-defined chalk ridge. Figure 2.7 shows that road and rail routes to Swanage make use of this gap today.

Perceptions of the natural environment

An example is the setting of a village within a particularly attractive rural landscape.

Travelling time/distance of a village or market town from an urban area

Activity 4

Study the map showing the situation of Elsenham in Essex (Figure 2.8). How do you think the perceptions of the natural environment and the travelling times to Bishop's Stortford, Cambridge and London might have affected the attitudes of Elsenham residents since the construction of the M11 and Stansted Airport?

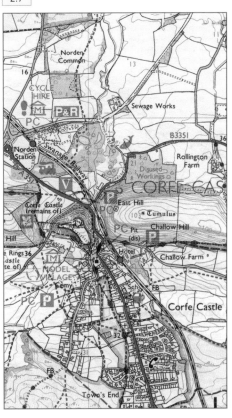

Figure 2.7 The situation of Corfe Castle

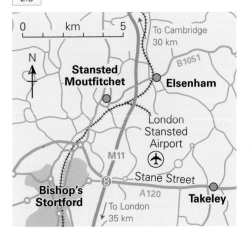

Figure 2.8 The situation of Elsenham

The juxtaposition of two distinctly different types of land use

An example is where an area of lowland arable farming gives way abruptly to upland pastoral farming.

Figure 2.9 | **The situation of Swaffham Prior and Swaffham Bulbeck, Cambridgeshire**

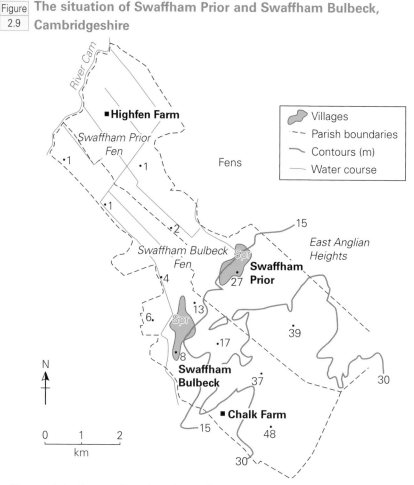

Figure 2.9 shows the situation of two villages of Anglo-Saxon origin in Cambridgeshire. Swaffham Prior and Swaffham Bulbeck are both situated at the junction of the dry chalk uplands (East Anglian Heights) and the wet peat lowlands (Fens). In the wider context of the region, as small ports since the Middle Ages, they were connected to The Wash. They were used by shallow draught vessels, which transported the fish and birds of the Fens, often in exchange for the crops and meat of the uplands. The two parishes are narrow elongated areas stretching from the uplands to the Fens. This shape provided each village with immediate access to resources: woodland; grazing land on the chalk of the drier uplands; as well as the peat, reeds and meadows of the wetter Fens.

Activity 5

Figure 2.10 **The situation of Warkworth, Northumberland**

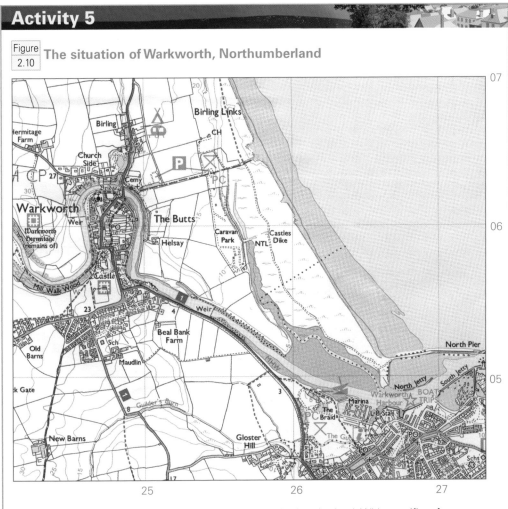

Study Figure 2.10, a 1:25 000 OS map of Warkworth, Northumberland. With specific reference to evidence from the map, identify and explain the advantages of Warkworth's situation.

Changing situation

Over time the situation of settlements can change. This can have a significant effect on their development. Situation influences the physical size, the total population, the population structure, service provision and other commercial activities. In fact, the entire status or importance of a rural settlement within the settlement hierarchy is influenced by its situation.

There are many examples of human and physical processes that cause rural settlements to experience a change in situation. The effect might cause either growth or decline in status.

Examples of these processes include:

- changes in communications networks, e.g. motorway construction, airport expansion, loss of a rail link
- planning decisions, e.g. planned growth of a nearby settlement such as a new town, redirection of industrial activity, designation of an area as a national park
- economic change, e.g. either depletion or exploitation of a natural resource, growth or decline of port activity in a region, investment in an office or retail park
- change in the physical environment, e.g. land reclamation, coastal erosion, alteration of a water course or drying of a lake

Case studies of changing situation

Kiveton Park

Kiveton Park is a village in South Yorkshire on the concealed Yorkshire, Nottinghamshire and Derbyshire coalfield. It is an example of a settlement whose situation in relation to coal deposits has changed due to technology and the increase in demand for coal. Until 1861, agriculture was the main occupation for its 300 residents. Then, in the 1860s, advances in mining technology opened up the concealed coalfield. By 1867, mine workers had been able to pierce the magnesian limestone in a deep shaft reaching the famous Barnsley Bed (or Top Hard seam of Nottinghamshire) at a depth of 400 m. This sudden access to new resources transformed Kiveton Park's situation. Ten years later, following the sinking of the first colliery shaft, its population had increased almost fivefold. After a century of mining, Kiveton Park's population peaked at 6300 in 1971.

Significant physical expansion accompanied population growth, along with an increase in shops and other services. Mining was the mainstay of employment until 1994, when the pit finally shut down with the loss of 1000 jobs. Since then, Kiveton Park's population has fallen. As a result, many services are no longer sustainable and have closed. This, together with unemployment, has added to the relative deprivation within the community.

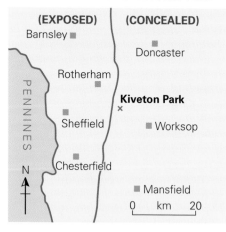

Figure 2.11 **The Yorkshire, Nottinghamshire and Derbyshire coalfield and the situation of Kiveton Park**

Wickham Market

In the decade 1971–81, the population of the rural market town of Wickham Market in Suffolk grew by 50%, from 1436 to 2154. The main reason was the construction of the A12 dual carriageway, which created improved accessibility both locally and within the regional setting (see Figures 2.12 and 2.13). Travelling time to the major centre of employment, Ipswich, and the expanding port of Felixstowe was reduced. Influxes of young working people, prepared to commute, and retirees caused significant changes in the population structure. The expansion of its primary school, old people's home and doctor's surgery can be explained by these changes. This was typical of many rural market towns affected by counterurbanisation.

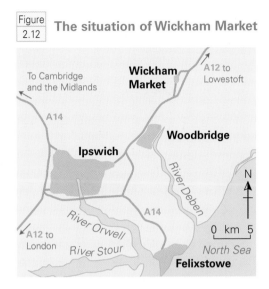

Figure 2.12 **The situation of Wickham Market**

Figure 2.13 **The A12 dual carriageway at Wickham Market**

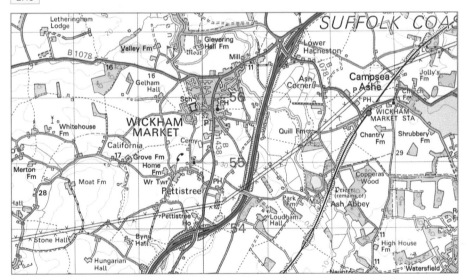

Rattenberg

An unusual example of changing situation is provided by the village of Rattenberg in Austria. Situated within the high and steep-sided mountains of the Tyrol range, the settlement is at the foot of a shaded, north-facing slope. It is claimed that many people suffer from depression because of the lack of sunlight in winter and that this explains the 10% loss of population between 2000 (514) and 2005 (467). Rattenberg's situation also affects numbers of tourists and the viability of local business. The EU-funded plan is to improve Rattenberg's situation by the use of computer-guided solar reflectors (heliostats). The intention is to provide sunlight for the community at ten hot spots (particular streets, attractive and historic buildings). It is hoped that this will reverse the village's economic decline and net migration loss (Figure 2.14).

Figure 2.14	The use of heliostats in the Tyrol Mountains (*The Times* World News, 27 October 2005)

Mirrors help villagers to come out of the shadows

By Roger Boyes

The residents of an Austrian village in the shadow of the Tyrol mountain range hope to beat the winter blues with dozens of giant mirrors.

They have backed a scheme to use the computer-controlled reflectors to bounce the rays of the sun around the mountains overlooking Rattenberg and onto their high street.

The initiative, backed by the European Union, could become a model for the whole alpine region. Whereas countries such as Slovenia market themselves as the "sunny side of the Alps", many communities are denied sunshine from November to February because they are on the wrong side of the mountain. They are being called the "valleys of despond".

Rattenberg's 460 inhabitants have unusually high rates of depression, according to Peter Erhard, the local doctor. He says that more of his patients are reporting sleeplessness, sadness, lethargy and poor self-esteem during winter.

The population has shrunk. "We have to keep our people here," Franz Wurzenrainer, the village mayor, said. "That is why we have launched this rather mad scheme."

The mirrors — or heliostats — can bounce rays to a target. The plan is to build a bank of 30 solar reflectors, 2.5m × 2.5m (8ft × 8ft), in the neighbouring village of Kramsach, 500 metres away on the sunny side of the Stadtberg mountain.

The mirrors tilt so that they can track the sun's position. The rays are sent through gaps in the mountains to a second bank of 30 mirrors in Rattenberg. The reflected sunshine will not be enough to bathe the whole village but it can create ten hot spots: chosen streets, historical facades, many of which date back to the 15th century.

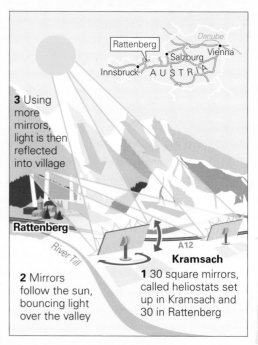

3 Using more mirrors, light is then reflected into village

Rattenberg

2 Mirrors follow the sun, bouncing light over the valley

1 30 square mirrors, called heliostats set up in Kramsach and 30 in Rattenberg

Kramsach

Depressed locals will be able to go on to the high street to tank up with sunlight.

The mirrors will cost about €2 million (£1.4 million), funded by the EU and the regional authorities, and should be operating by the first half of 2007.

Although the principle of the heliostat was developed as early 1865 by Jean Bernard Foucault, it has never been deployed to light a community.

Markus Peskoller, the head of the Bartenbach Light Laboratories in Austria, is confident that the mirrors are a solution for the Alps. "We can help the inhabitants of other mountain villages to see the light," he said.

Morphology

Morphology is the shape or form of a rural settlement evident from a map or aerial photograph. Features of the physical environment and human factors such as communications, the social and economic requirements of particular cultures, including the degree of communality, and modern planning are all factors that can influence the morphology of an individual rural settlement.

Many attempts have been made to classify rural settlement forms. The main difficulty is the great variety of village shapes and the fact that many rural settlements have changed their form throughout their history. Nonetheless, three main types are recognised: **linear**, **nucleated** and **dispersed**.

In addition to the overall shape, the morphology of a rural settlement includes its internal street layout, buildings and open spaces. Comparison of the latest OS map with earlier editions may reveal the pattern of development of a village over the past 100 years or so.

Activity 6

Study the rural settlements on a 1:50 000 or 1:25 000 OS map of an area near your home.

(a) Draw up a table giving the names and shapes of five villages.

(b) Using map evidence, suggest reasons for these village shapes, such as aspects of relief or communications.

(c) Use map evidence to help explain any exceptions or modifications to the basic shapes.

(d) Describe the layout of streets, buildings and open space in one of the larger villages.

(e) State the evidence for any changes in the layout of your chosen village through time. Suggest reasons.

Linear settlements

Villages are most commonly linear in shape. Linear settlements owe their form to a variety of human and physical factors:

- Man-made dykes and natural levées create long, narrow, dry sites in areas liable to flooding, e.g. North Muskham or Carlton-on-Trent in Nottinghamshire.
- Roads, rivers and canals provide good accessibility where trade was important for all households, e.g. Long Melford, Suffolk.
- The agricultural organisation of the original settlers, especially where land was allotted in long rectilinear holdings running back from a narrow frontage on a road or waterway, e.g. Barroway Drove, Norfolk.
- Narrow, steep-sided valleys offered limited space for building along the valley, e.g. Cwm-parc, Rhondda Cynon Taff or Lochranza, Isle of Arran.

- A well-defined break of slope, or the junction between two rock types, where strip farming allowed each family access to the better soils or the opportunity to exploit more than one local environment, e.g. Staxton, North Yorkshire.
- Early planning/control by manorial lords, e.g. Helmingham, Suffolk.

Activity 7

Figure 2.15 **Cwm Penmachno**

D. Barker

Afon Machno

Study Figure 2.15, a photograph of the village of Cwm Penmachno in North Wales. Using evidence from the photograph, state three reasons for the linear form of this settlement.

Case study: a linear settlement

Long Melford in west Suffolk was a classic example of a linear village. Its original shape was ribbon-like, following a Roman road. Immediate access to this line of communication was essential for every householder for trade when the cloth-weaving industry became as important an economic activity as farming in the fifteenth and sixteenth centuries. The steeper slopes on its eastern side plus the marshes of the Stour floodplain and its tributary Chad Brook to the west prevented early construction, helping to maintain the linear shape.

The building of two mansions, Melford Hall and Melford Place, at each end of the village further accentuated its linearity, since estate cottages for employees such as servants and grooms were built nearby. In addition, the railway station near the railway junction attracted agricultural industries to the southern end of the village in the nineteenth century.

Following a period of stability with relatively little new building, modification of the original linear shape has occurred, but only since 1970. In fact, the population of Long Melford was in decline between 1928 (2560) and 1961 (2375), but a significant increase of 45% between 1961 and 2001 (3433) explains the recent expansion eastwards towards the dismantled railway line (see Figure 2.16).

Figure 2.16 **The morphology of Long Melford in (a) 1928 and (b) 2005**

(a)

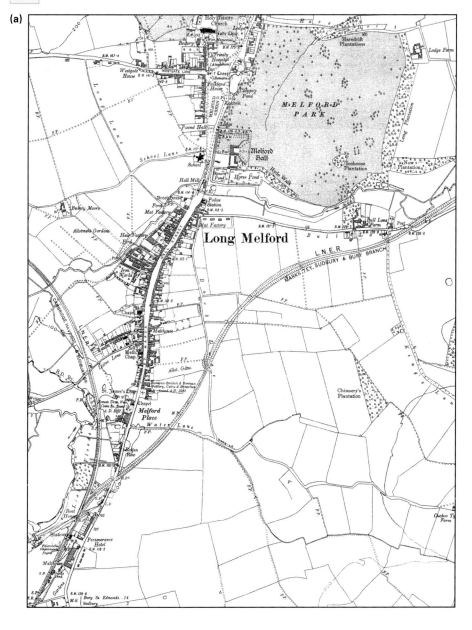

(b)

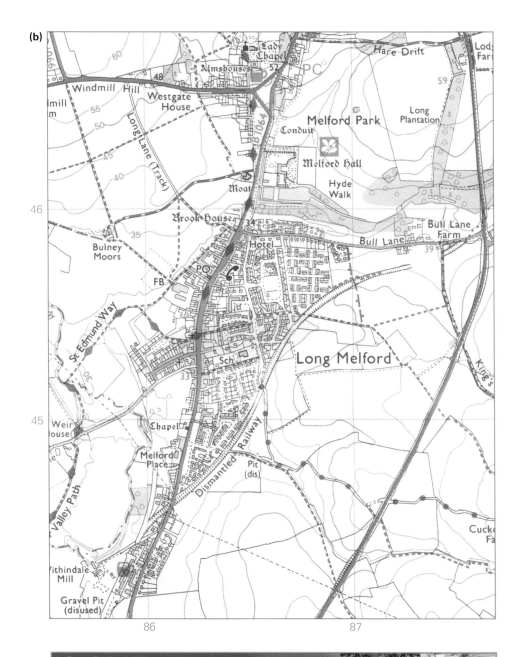

Activity 8

Study the 1928 OS map in Figure 2.16. With specific reference to map evidence:

(a) What is the distance from the railway station to Holy Trinity Church?

(b) What types of employment were there in this village in 1928?

(c) Suggest reasons for the greater road width in the central part of the village.

Nucleated settlements

Nucleated settlements are compact, with buildings clustered around or near to an important central feature. This process of clustering may be the result of environmental, social or economic factors, including planning policies. These include:

- the open field system of farming developed by Anglo-Saxon settlers in southern England — under this system, access to the fields and the adoption of communal farming practices were best achieved in a centrally located nucleated settlement (see Figures 2.17 and 2.18)
- the use of the village green (often with pond and church) for livestock security, and space for communal activities such as fairs and other social events
- a point of route convergence, such as a crossroads or bridging point, where a higher degree of accessibility enabled trade and the development of commercial activity
- access to a reliable water supply in an otherwise dry area, for example where limestone is the dominant rock type
- avoidance of marshy ground or flooding on an elevated dry-point, such as a well-drained river terrace or an alluvial fan of sand and gravel
- proximity to a place of employment based on a mineral resource such as a colliery
- where rural development has been planned and polarised at a key settlement

Figure 2.17 **The morphology of Bradworthy, Devon**

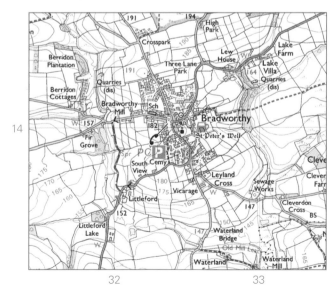

Figure 2.18 The morphology of Old Buckenham, Norfolk

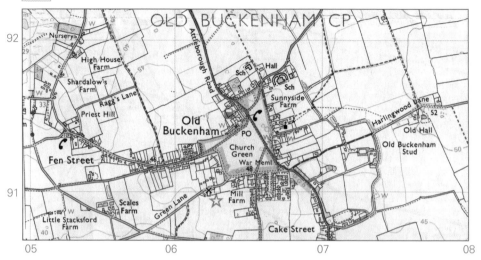

Activity 9

Figures 2.17 and 2.18 are 1:25 000 OS extracts of two nucleated villages: Bradworthy in Devon and Old Buckenham in Norfolk.

(a) With reference to the evidence of the OS maps, identify the physical and human factors that have led to nucleation in each instance.

(b) Many villages have their own websites. Use www.bradworthy.co.uk to investigate reasons for the growth and development of this village as a nucleated settlement.

Dispersed settlements

Many villages have a more dispersed layout, often with one slightly larger hamlet plus two or three others, all within the same parish. The houses are scattered and joined by a network of lanes with no obvious centre.

The process of dispersal that led to this form of settlement often occurred over a long period of time. Dispersal was closely associated with changing agricultural organisation and the increasing demand for additional farmland. Land was reclaimed from marsh and wasteland, woodland was cleared, moors and even common lands were used to create new farms. This led to farmsteads and smaller hamlets being constructed away from the original hamlet.

A similar process occurred as a result of enclosure. Following Parliamentary Enclosure in the eighteenth and nineteenth centuries, new land holdings were consolidated and new farm dwellings and hamlets were built, again away from the original settlement.

Examples of dispersed settlements

Wadshelf

Activity 10

Figure 2.19 **Dispersal of farmsteads resulting from field enclosure and use of the moors west of Wadshelf, Derbyshire**

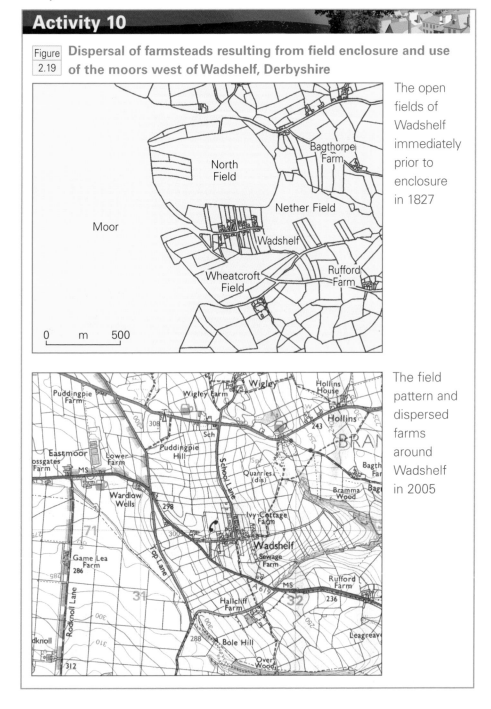

The open fields of Wadshelf immediately prior to enclosure in 1827

The field pattern and dispersed farms around Wadshelf in 2005

Activity 10 (continued)

Study the two maps of the area around Wadshelf in Derbyshire (Figure 2.19). Wadshelf is located 5 km west of Chesterfield on the edge of the Peak District moors.

Describe and explain the changes in **(a)** the field patterns and **(b)** the distribution of settlement in the area shown.

Wainfleet St Mary

Map evidence shows that the building of dykes to the southeast of Wainfleet St Mary in Lincolnshire, progressively over time and in lines parallel to the coast, has created a landscape that has gradually been reclaimed for farming. The pattern of roads — long, straight and perpendicular to the old coastline — and the location and names of the farms suggests colonisation of the land eastwards in a dispersed pattern.

| Figure 2.20 | Dispersal of farms resulting from reclamation of marshlands southeast of Wainfleet St Mary, Lincolnshire |

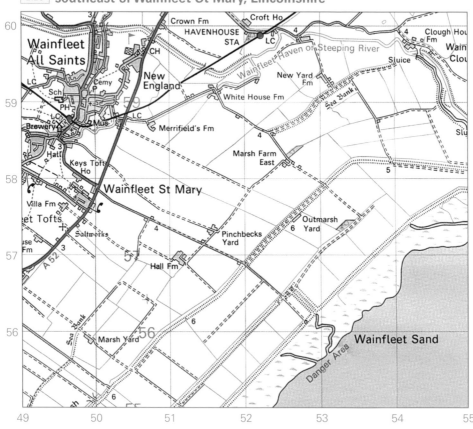

Saxtead

Place name evidence is often indicative of dispersal. Elements such as **Green**, **Little Green**, **End** or **Row** may be added to the basic settlement name. Examples can be seen in Figure 2.21. The moated farmhouse is an additional feature of the landscape in this area around Saxtead in Suffolk. Status and prestige rather than defence is the current interpretation of these features. What do you think were the other advantages of creating moats around farms?

| Figure 2.21 | **A dispersed village in the parish of Saxtead** |

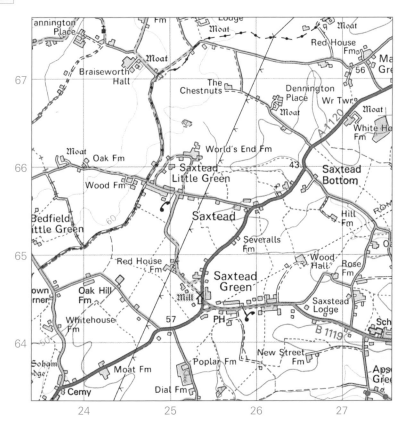

Settlement patterns

The terms nucleated settlement and dispersed settlement are used to describe the morphology of individual settlements. They should not be confused with **nucleated settlement pattern** and **dispersed settlement pattern**, which refer to the overall spatial pattern of settlements in a particular region or area.

Nucleated settlement patterns

A nucleated settlement pattern is where the spatial distribution of rural settlements in a region is dominated by villages. There are often scattered farmsteads between the villages. This pattern tends to develop in lowland areas, where relief, climate and soils encourage farming systems that support higher population densities.

Many of the villages in eastern and southern England are part of the nucleated settlement patterns resulting from Anglo-Saxon colonisation. Figure 2.22 shows a well-organised nucleated settlement pattern in the Yorkshire Wolds near Bridlington. The benefits of communality in farming could be derived in this area of rich soils, gentle slopes and equable climate. Access to points of water supply on the chalk, access to land, sharing of labour, equipment and other resources, and protection of livestock were all made easier for the groups of farmers in a nucleated settlement pattern.

Other possible reasons for the development of nucleated settlement patterns include the need for defence in pioneer areas, a multiplicity of wet-point and/or dry-point sites in an area, planning policy within a region, and the establishment of pit villages on a concealed coalfield. In many areas, the current settlement pattern is the product of more than one of these physical and cultural influences.

Figure 2.22 **A nucleated settlement pattern on the Yorkshire Wolds**

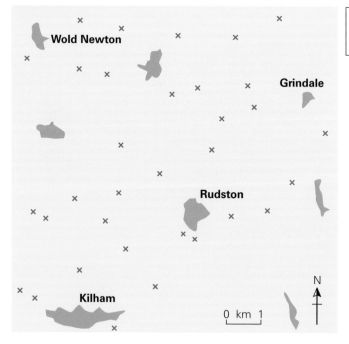

Dispersed settlement patterns

A dispersed settlement pattern is where the spatial distribution of rural settlements in a region is dominated by isolated farms. There are few villages except in the more sheltered and accessible valley locations.

This is often, but not exclusively, a feature of upland areas, where the natural resources for agriculture are poor. The pastoral farming that has developed is characterised by large farms and the farmsteads are widely spaced since the farmers need to be relatively near their livestock. Figure 2.23 shows the dispersed pattern of settlements in an upland region of the Pennines near Malham in North Yorkshire.

Figure 2.23 **A dispersed settlement pattern in the Pennines near Malham, North Yorkshire**

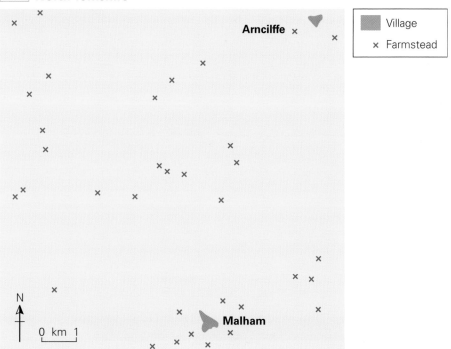

In addition to physical factors, cultural and social factors have contributed to dispersed patterns of settlement. As the need for defence declined and freedom from enforced communality in the lowlands under the feudal system became possible, settlers moved to the uplands and to newly reclaimed areas of land. The advantages derived from owning their own land were sufficient incentive to move away from the village.

You should be aware that not all dispersed settlement patterns are found in upland areas; they are also a feature of many fenland areas. Figure 2.24 illustrates the effects of government planning and developments in technology on the distribution of farmsteads in an area below sea level. In the Dutch landscape, Kampereiland has an irregular, dispersed pattern of farmsteads, which developed as these older polders were reclaimed piecemeal. On the newer Northeast Polder, the farmsteads are located at regular intervals around the village of Ens in a planned, regimented landscape of intensive arable farming.

| Figure 2.24 | A dispersed settlement pattern on the Dutch polders |

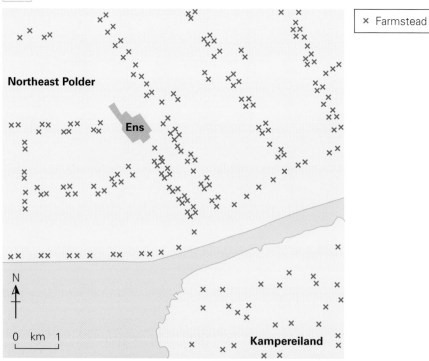

Activity 11

Study Figure 2.25, which shows the pattern of settlement in part of north Wales, and Figure 2.26, a photograph of the area indicated on the map, northeast of Tryfan.

(a) Using the information shown on the map (Figure 2.25), describe the pattern of settlement.

(b) State evidence shown in the photograph (Figure 2.26) to suggest reasons for this settlement pattern.

Activity 11 (continued)

Figure 2.25   **The settlement pattern in Snowdonia**

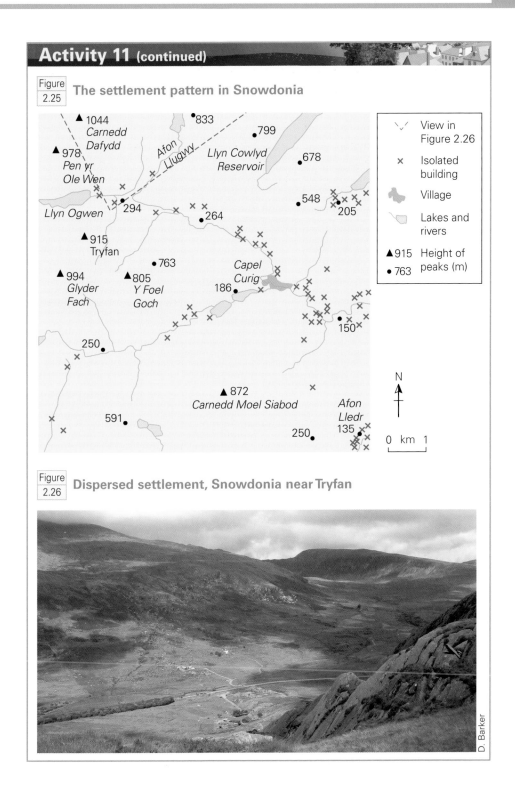

▲1044 Carnedd Dafydd
•833
•799
▲978 Pen yr Ole Wen
Afon Llugwy
Llyn Cowlyd Reservoir
•678
•548
•205
Llyn Ogwen
294
•264
▲915 Tryfan
•763
Capel Curig
▲994 Glyder Fach
▲805 Y Foel Goch
186•
150•
250•
250•
591•
▲872 Carnedd Moel Siabod
Afon Lledr
135•

Legend:
\/ View in Figure 2.26
× Isolated building
Village
Lakes and rivers
▲915 Height of peaks (m)
•763

N

0 km 1

Figure 2.26   **Dispersed settlement, Snowdonia near Tryfan**

D. Barker

Evolution of rural settlement patterns

The current settlement pattern in nearly all regions of the UK has evolved over many centuries. During this time a variety of influences — physical, economic, social, political and cultural — have left their imprint on individual rural settlements and on the wider settlement pattern of the area. As a result, it is important to note that many settlement patterns are 'composite', containing both nucleated and dispersed elements — for example, areas affected by eighteenth-century enclosure.

The examples referred to in this chapter so far show that detailed analysis of Ordnance Survey maps, especially at 1:50 000 and 1:25 000 scales, can help in the identification of these factors. Relief, altitude, drainage, water supply, and field and parish boundaries provide information from which it is possible to deduce the main features of the site, situation, agricultural organisation and economic activity of most rural settlements.

The significance of these factors is particularly apparent when viewed in a historical context. Place names and past structures shown on maps can be dated or assigned to a period of time and colonisation. This evidence suggests distinct periods in which invading groups have left evidence of their occupation (see Table 2.1).

Table 2.1 Evidence of stages in the evolution of rural settlement patterns available from OS maps

Period	Place name	Meaning/example	Evidence of settlement
Neolithic (c. 4000–2000 BC)	–	East Heslerton Brow	Long barrows (burial mounds)
Bronze Age (c. 2000–800 BC)	–	Upland area/Roughtor, Bodmin Moor	Round barrows Hut circles Field systems Stone circles
Iron Age (c. 800 BC–AD 71)	–	Iron Age villages in west Cornwall near Chysauster and Porthmeor	Hilltop camps and forts Earthworks Celtic fields Lynchets
Roman (c. AD 71–410)	...caster, ...cester, ...chester	Fort/Brancaster, Norfolk	Roman roads Sites of villas Signal stations

Period	Place name	Meaning/example	Evidence of settlement
Anglo-Saxon (c. AD 410)	Primary settlement ...ham, ...ing, ...ingham ...ton ...borough, ...burgh ...bridge, ...ford Daughter settlement ...ley ...stead ...stow ...wick Woodland clearing ...den ...hurst Fen names ...ey, ...ea	Homestead/territory of... Enclosure with dwellings Fortified place Bridge, ford Clearing in woodland Place/position/site Holy place Outlying hut/dairy farm Pasture for swine in wood Copse Island in marshy area	Tumuli (burial mounds) Ship burials Woodland clearance Nucleated settlement pattern
Scandinavian (–1066)	...by ...ey ...thorpe ...toft ...kirk	Village/homestead Island Hamlet/daughter settlement Homestead/clearing Church	–
Medieval (c. 1066–1540)	–	Leiston Abbey, Suffolk Wykeham Abbey, North Yorkshire Framlingham, Suffolk Moat Hall Farm and fishponds, Parham, Suffolk Swaffham, Dereham and Fakenham, Norfolk Lower Ditchford, Gloucestershire	Religious sites, monasteries, priories Castles Manor houses, halls Fishponds Moated farmhouses Emergence of small market towns Village desertions
Early modern (c. 1600–1800)	–	Kilby, Leicestershire Holkham and Felbrigg, Norfolk	Parliamentary Enclosure, field boundaries Dispersed settlement patterns
Modern (c. 1800 to present)	–	Wadshelf, Derbyshire: enclosure of moorland	Parliamentary Enclosure Influence of canals, roads, railways, motorways Village growth/decline Evidence of planning, key settlement policy, airports, industrial estates

Activity 12

Try adding to Table 2.1 through your own research. Look for named examples of evidence for settlement on the OS map of your home area, or any other area, since the Neolithic period.

Case study: the Vale of Pickering, North Yorkshire

Most of the influences on settlements and settlement patterns outlined in this chapter can be illustrated in the Vale of Pickering and its adjoining uplands. Figure 2.27 shows part of the Vale of Pickering known as The Carrs. Settlements in this area are concentrated along the line of the two A roads, which follow the break of slope between the uplands and the floor of the vale.

Figure 2.27 **The Carrs, Vale of Pickering**

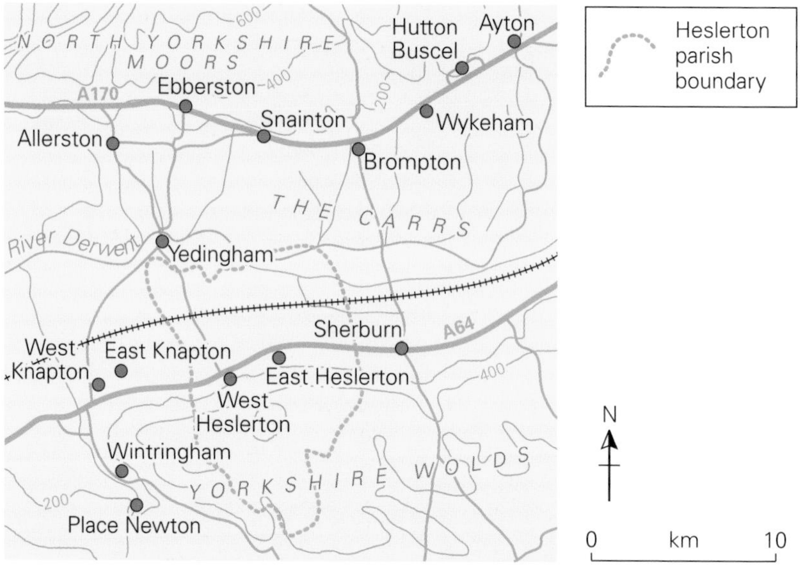

The Vale of Pickering is an area of flat or gently undulating land between the North Yorkshire Moors and the escarpment of the Yorkshire Wolds to the south. On the northern side of the vale, between Allerston and Ayton, the nucleated settlements are close together. They are located above the level of flooding of the River Derwent, which drains the area; they all have access to a water supply from the springs that emerge from the limestone of the North Yorkshire Moors; and they are at the junction of the differing soil types and environments for farming between upland and vale.

The same factors operate on the southern boundary, where the chalk of the Yorkshire Wolds gives way to the clays and peats of the vale. Settlements such as East and West Knapton, East and West Heslerton, and Sherburn are classic spring-line settlements of Anglo-Saxon origin.

On the floor of the vale, the settlement pattern is one of dispersed farmhouses. This area was settled much later, following drainage and enclosure in the seventeenth and eighteenth centuries. The only village here is that of Yedingham, at a significant bridging point of the Derwent.

The parish of Heslerton

A closer study of the parish of Heslerton illustrates the wide variety of factors that have influenced the settlement pattern within the Vale of Pickering.

Figure 2.28 is a cross-section north to south through this parish. The annotations are based on deductions made from the evidence of the 1:25 000 OS Explorer map. Although the record of OS maps might be incomplete, it is possible, by studying an OS map of 1:25 000 scale, to establish the physical factors and the cultural or social influences involved in the evolution of rural settlement patterns in most areas of the UK.

Figure 2.28   **Influences on settlement in the parish of Heslerton**

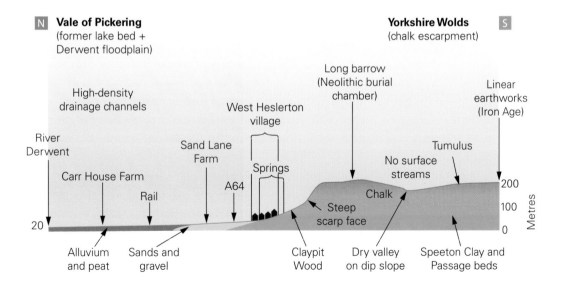

The boundary (Figure 2.27) shows that the parish of Heslerton lies to the south of the Derwent and it includes land on both the Wolds and The Carrs. The Wolds are completely devoid of surface drainage and their smooth, rounded contours at heights of approximately 150–190 m suggest a porous chalk landscape. There is a dendritic network of dry valleys on the dip slope.

The Carrs are lower, at about 22 m, and flat. The high density of surface drainage channels is typical of the alluvial or peaty deposits found in many former marshy fenland areas. There is a gentle concave slope at the foot of the scarp face on which the main settlements have been built. Names such as Claypit Wood and Sand Lane Farm give evidence of the main rock types found in this zone (the Speeton Clay and Passage Beds beneath the chalk).

The **site** of the main village, West Heslerton, is on this gentle lower slope at the foot of the scarp face. Fresh water is supplied where springs emerge at the junctions of the clay (impervious), sand and chalk (porous). Early perception of the potential flood hazard of the Derwent was another reason for the more elevated site of the original village, which lies between the 45 m and 55 m contours. These contours are widely spaced, suggesting relatively flat ground, easier for construction of dwellings.

The **situation** of West Heslerton can be explained by its proximity to a wide variety of resources nearby. It had immediate access to the drier upland used for the grazing of livestock, and timber was available on the steep scarp face often left uncleared. Chalk, sand and clay were easily extracted from the local rock outcrops. The lowland marshes of the Derwent floodplain could be used for meadowland and, where drier, crop production; the streams and ponds provided a fish supply.

There is evidence of prehistoric settlement in the parish as shown by the Neolithic long barrow (burial mound) on top of the chalk escarpment above East Heslerton Brow (OS GR 938753). Linear earthworks from the Iron Age (of which at least 3 km remain visible in this parish alone) were probably designed to prevent livestock from roaming, rather than for any defensive purposes.

There is no map evidence of Roman occupation but it is thought that the main road was an access route from York via Malton to the Roman signal station at Scarborough.

Anglo-Saxon occupation is more obvious in the place name ending …ton. Hesler**ton** was a primary settlement or enclosure near a hazel wood. The nucleated shape of the village and the nucleated settlement pattern along the spring line can be linked to the communal farming systems that developed during this period of Anglo-Saxon colonisation.

Being near to the east coast of northern England, Scandinavian influence on the settlement pattern in the Vale of Pickering was strong. There is no evidence in this specific locality but elsewhere in the vale, place name endings such as Gris**thorpe**, Osgood**by** and Oswald**kirk** are all clear evidence of this phase.

The dispersed settlement of The Carrs came much later. This may have coincided with dispersal of farmsteads following Parliamentary Enclosure in the eighteenth and nineteenth centuries but this was also dependent on the improving technology and ability at that time to reclaim marshland. The dispersal outwards from the main spring line settlements is evident in the straight minor roads and tracks, which extend from the A64 towards the River Derwent. Farm names such as Carr House Farm and East Heslerton Carr House indicate this later dispersal onto the marshy areas or carrs.

Modern influences can be seen in the buildings along both the A64 and the Malton to Scarborough rail connection, which runs along the floor of the vale.

Activity 13

Search the websites for **(a)** the 2001 Census Key Statistics for the Parish of Heslerton and **(b)** the Genuki West Heslerton Directory of Trades and Professions for 1905, to investigate changes in employment structure between these two dates.

3 The rural settlement hierarchy

Settlement hierarchies are the rank order of importance of settlements in an area. The related concepts of **central place**, **catchment area**, **threshold population**, **range** and **centrality** are important in understanding the characteristics of rural settlement hierarchies. They help to explain the rank order of importance of settlements (the functional hierarchy) as well as their spatial distribution in a rural area.

In the first part of this chapter, these concepts — which underpin all settlement hierarchies — are explained. Towards the end of the chapter the characteristics of settlement hierarchies are examined in terms of both theoretical models and case studies.

Settlement hierarchies are not static: they are systems that are subject to change through time. For example, there may be changes in the population and the number and type of shops and services of individual settlements. As a result, the sustainability of a settlement or the entire settlement hierarchy may be affected. Changing levels of importance or status of settlements within the hierarchy and the issues of sustainability are explained in Chapter 4.

Concepts

Central place

Central places are settlements that provide services for their own resident population and for people in the surrounding area. Within the rural settlement hierarchy, of which they are part, these central places vary in size and importance, ranging from market towns to small villages. They may be categorised as relatively high-, medium- or low-order centres according to the number and type of functions found within them.

By definition, there are countless central places of all types in all rural societies. Each one is an individual service centre with its own characteristics. Collectively they form a wider hierarchical pattern of settlements within a particular rural region. This organisation of service provision has evolved over a long period of time to meet the needs of the population.

At the global scale, central places are diverse. They are closely linked to the natural environment, to the rural economy and to the cultural and political geography of the area in which they are situated. Consequently they may have quite different characteristics.

Case studies of central places

Hualcayan, Ancash, Peru

Hualcayan is a remote village of 180 people in the Cordillera Blanca of northern Peru (see Figure 3.1). The inhabitants depend largely on locally grown potatoes and grain plus pig and guinea pig rearing. Other items are obtained from Caraz, the nearest town, which is a 2-hour drive away along steep and winding dirt tracks. The only goods for sale in Hualcayan are bottled drinks, canned fruit, sweets and biscuits. These are available in five households, one of which has a rudimentary café (there is an occasional passing trade generated by tourism in the Huascaran National Park). There is a truck service to Caraz once a week. The most important service in the village is its school; this serves the village itself and the surrounding isolated farms. There are 70 pupils in two classrooms — one for 9–16-year-olds and one for the under 9s.

| Figure 3.1 | **Hualcayan, a remote central place in the Andes** |

Lizzie Barker

Richmond, North Yorkshire

Richmond is a long-established market town with a population of over 8000 people. It lies between the Pennines and the Vale of York in North Yorkshire. It is highly accessible — at the hub of routes from the uplands and the vale, with frequent bus services. The town serves a relatively high-density, mobile rural population living in many surrounding villages. Richmond's wide range of local businesses, with over 100 shops and services, can be identified on: www.richmond.org.uk. Darlington, the nearest large town, is only 15 km along the A1(M).

Hualcayan and Richmond are in countries with different physical relief, levels of economic development, and communications. As a result, the settlements have different status. Nevertheless, their size, distance from other centres and importance are governed by the same basic concepts of threshold and range, and they are both part of a wider settlement hierarchy.

Catchment area

In any rural region there is a strong relationship between the level of service provision in a central place and the surrounding area over which it exerts an influence. Often referred to as its **catchment area**, this area is accessible and functionally linked to the central place. The people living within it interact with the central place and often develop an allegiance to it for certain services.

A primary school, for example, located in a rural market town is usually attended by the children of primary age in that town plus those of the smaller neighbouring villages, hamlets and individual farmsteads. The geographical extent of the school's catchment area can be mapped by encircling the home locations of the pupils, as shown in Figure 3.2.

Figure 3.2 **The catchment area of Sir Robert Hitcham's Primary School, Framlingham**

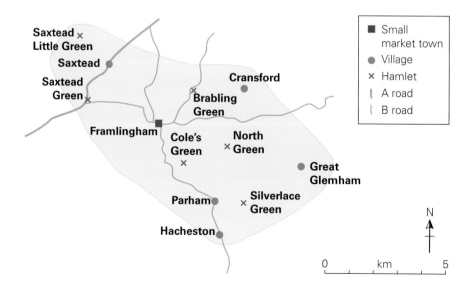

The term catchment area is frequently used for schools. For other functions, the same concept is also known as the settlement's **sphere of influence, trade area, hinterland** or **interaction field**. In this chapter, 'catchment area' is used in its generic context. The delimitation of a catchment area can be based on one individual function, such as a school or a particular shop, or it can be more composite (see Figure 3.3), depending on the purpose of the investigation.

The size and shape of any catchment area is influenced by a number of human and physical factors. These include:
- the number of shops and services
- the type of shops and services — proportion of high/low order
- the range of functions available
- the proximity of similar competing centres
- accessibility — road networks/public transport
- travelling times
- population densities in the surrounding area
- affluence of the population
- physical barriers, e.g. estuary, mountains

Activity 1

| Figure 3.3 | The catchment area of Richmond, North Yorkshire, for convenience goods and services |

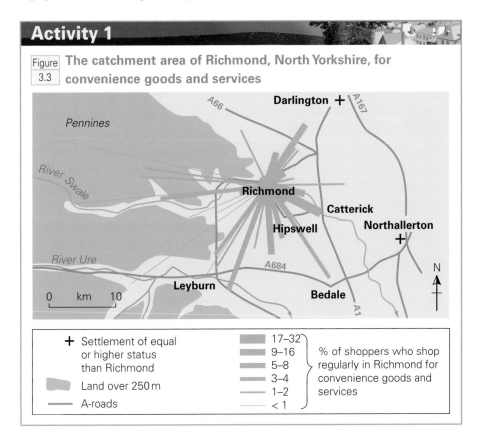

Activity 1 (continued)

Figure 3.3 shows the catchment area of Richmond in North Yorkshire. Richmond is a market town and the desire lines are based on the percentage of shoppers who shop regularly for convenience goods and services. With specific reference to Figure 3.3:

(a) describe the size and shape of the catchment area

(b) suggest reasons for its irregular shape

Techniques for measuring and estimating catchment areas

The area over which a rural settlement exerts an influence can be measured by field investigation. This is useful in the study of rural settlement hierarchies because it allows comparison of catchment areas at different levels — for example, villages and market towns, low-order and high-order goods and services.

Fieldwork based in the central place itself usually depends on the researcher being able to locate the places of residence of:

- regular customers visiting a particular shop or service
- customers receiving goods delivered by a particular store
- daily commuters to a market town
- pupils of primary and secondary school age

Individual indices such as these are relatively simple to survey and record. You may be able to suggest others that are equally useful and effective.

Using these kinds of data, it is possible to map and compare the catchments of individual functions, for example at the grocery, hardware or clothing levels. Although sometimes harder to ascertain, catchment areas of higher order services such as local libraries, solicitors, dentists and secondary schools can also be established.

Field investigation in the surrounding countryside, rather than in the central place itself, is another way of establishing the catchment area of a rural settlement. An exponent of this technique was H. E. Bracey, who studied north Somerset settlements in 1953. One piece of research was based on the rural areas around the towns of Bridgwater and Weston-super-Mare; this could be adapted for any rural area or rural settlement.

Bracey established 15 services that were thought to be basic requirements, including clothing shops, household goods, medical services and financial services. Selected villagers, such as teachers, clergymen and parish council chairmen, were asked where they obtained each service. If the villagers regularly used only one central place for a particular service, one point was scored by the village;

each village could therefore score a maximum of 15 points. By plotting this information on a map, it was possible to show differing degrees of intensity of the trade area of central places by drawing isolines (see Figure 3.4).

Figure 3.4 **The catchment area of Bridgwater according to H. E. Bracey**

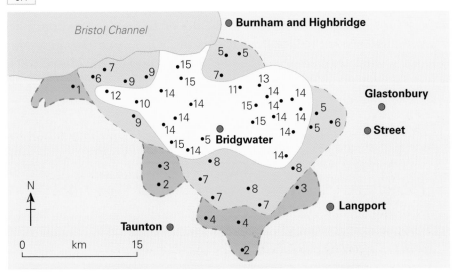

Activity 2

(a) What are the advantages and disadvantages of Bracey's technique?

(b) What changes have taken place in rural areas since 1953 that would make the technique more difficult to apply and more unreliable today?

(c) Design a questionnaire that you could use in the surrounding villages and hamlets to establish the catchment area of a small market town. Which functions would be the most useful indices? How would you establish reasons for the patterns found?

Catchment areas can also be predicted using theoretical models. These models include the work of W. J. Reilly. Reilly's model is relatively simple. If there are two settlements, A (larger) and B (smaller), the distance of the break point (theoretical catchment boundary) from the smaller settlement (B) is derived from the equation:

$$\text{distance of the break point from the smaller settlement B} = \frac{\text{distance in km between settlement A and settlement B}}{1 + \sqrt{\dfrac{\text{population of settlement A}}{\text{population of settlement B}}}}$$

This model is based on the following assumptions:

- The main function of the settlement is a service centre.
- The trade area is proportional to the population of the central place. (The total number of shops and services could be used in the equation instead of population. Why might this be a more reliable modification? Are there any other useful alternative indices that could be used?)
- The trade area is proportional to the distance of the central place from other similar neighbouring centres. (Journey time might be a more suitable measure today — why? Are there any other alternatives to the measure of distance?)
- Population densities in surrounding rural areas are all the same and there are no variations in wealth, demand and accessibility.

This technique of calculating the break point between two settlements could be extended to generate the entire theoretical catchment of a central place. In the case of the small market town of Debenham, for example, it is possible to map the predicted catchment by establishing its break point for each neighbouring centre, as shown in Figure 3.5.

Figure 3.5 **The theoretical catchment of Debenham according to Reilly's break point formula**

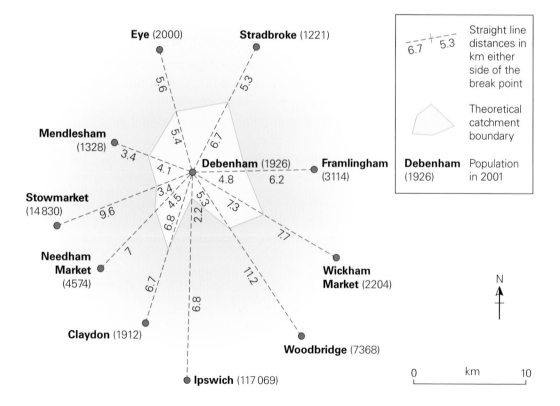

Activity 3

(a) For your extended study, how would you conduct a personal investigation to identify and explain the differences between the theoretical (predicted) and the observed (real) catchment of a small market town?

(b) What would be the difficulties in establishing the theoretical catchment?

(c) How would you identify and represent the observed catchment area(s)?

(d) What methodology would you adopt to help explain the differences between theoretical and observed patterns?

Threshold population and the range of a good or service

The concepts of threshold and range explain the location of the different types and numbers of shops and services within rural settlement hierarchies. **Threshold population** is defined as the minimum number of people (or expenditure) required to maintain the economic viability of a shop or service. The **range of a good or service** is the maximum distance customers are prepared to travel to purchase it. Do not confuse use of the term *range* in this context with the range or variety of goods and services found within a particular settlement.

High-order goods include durable items such as cars, washing machines and three-piece suites; high-order services include solicitors, schools, libraries and hospitals. Since items of furniture, for example, are high cost and are bought relatively infrequently, people are prepared to travel longer distances to obtain them. They travel in order to get their exact requirements, to compare prices and to make a choice. Similarly, people are prepared to travel greater distances to secure the services of a solicitor. Higher-order functions such as these therefore have a greater range. To be economically viable, a furniture store or school must locate where its threshold population can be met. This is usually in the larger central places, which are more accessible.

Low-order goods are convenience items such as bread and groceries. These are found in most rural settlements, even small villages. Often they are combined in one outlet, such as a village store or post office. Convenience goods such as these are consumed quickly and regularly, and therefore they are bought frequently, perhaps every day. Many potential customers are not prepared to travel long distances just to buy these items alone, but they will travel short distances to a local shop for them. Low-order goods therefore have a short range and the shops that sell them are found in the smallest central places, since even here their threshold populations can be met. The causes and effects of changes to this traditional pattern are discussed in Chapter 4.

Centrality

The term **centrality** is defined as the importance or status of a central place within the settlement hierarchy. It is therefore a measure of a settlement relative to others within the same hierarchy. The settlement with the most services and which dominates at the top of the hierarchy is said to have the highest degree of centrality; smaller settlements with fewer services have lower degrees of centrality.

Centrality within a particular hierarchy can be measured by any of the following:
- number of shops and services
- type of shops and services
- total retail floor space
- retail turnover
- numbers employed in shops and services of different types
- size of catchment area

These are all significant measures, since they relate directly to the functional importance of the central place. Other indices, such as network connectivity (i.e. the extent to which places in a transport network are connected directly to each other), car park spaces, total population and even size of built-up area are useful, since they are also linked to the service provision of a particular settlement.

We should not confuse the concept of centrality with the idea of a geographically central location. Although most settlements of importance within a hierarchy do tend to be more centrally located, it is the accessibility to the smaller villages, hamlets and potential customers in the surrounding area that is the key factor in their status as a service centre.

Activity 4

A simple, effective measure of the status of settlements has been devised by W. K. D. Davies:

1 Calculate the coefficient of location for a particular function in settlement A.

$$\text{coefficient of location of a function} = \frac{\text{number of outlets of the function in settlement A}}{\text{total number of outlets of the function in the hierarchy}} \times 100$$

2 Add the coefficients of all the functions in settlement A to obtain a total score for that settlement.

Activity 4 (continued)

3 Calculate the total score for each settlement to find the rank order of importance. The settlement with the highest total score has the greatest degree of centrality within that particular hierarchy.

You are given the task of establishing the rank order of centrality of the settlements in an area that shows rural settlement hierarchy.

(a) How would you collect the relevant data? What would be the difficulties in doing so?

(b) Can you suggest any refinements to the Davies index of centrality?

The settlement hierarchy

A **settlement hierarchy** is the rank order of importance of towns, villages and hamlets in an area. The rank order is usually based on a measure or index of the settlements' functions, such as the number and the type of shops and services, total retail floor space or centrality index.

The settlement hierarchy concept can be applied at any scale, from global and national to regional and local. In most rural areas the hierarchical pattern of settlements has developed over a long period of time and it is the most efficient way in which shops and services are located to meet the needs of the population.

The relationships between threshold, range and catchment area determine the size and spacing of central places in the functional hierarchy. The geographical spacing of central places depends, to a large extent, on the distance or time that customers are prepared to travel to obtain their needs. Distances travelled tend to be greater for higher-order goods and services. Hence market towns that provide higher-order functions have larger catchment areas and are further apart than villages that provide only low-order functions. As a general rule, the larger the central place, the larger its catchment area.

Most settlement hierarchies include a number of different levels or categories of settlement. There are many small settlements (villages and hamlets) at the lowest level of service provision and only a few large centres (towns and cities) at the highest level.

At the top of the hierarchy in a rural region are the largest central places. These are few in number and they are widely spaced apart. Each offers a wide variety of shops and services, has an extensive catchment area and the highest centrality. At the regional scale, in the rural region of East Anglia, high-order goods and services are found in large central places such as the county towns of Norwich and Ipswich, which are 65 km apart. Department stores are typical high-order functions, selling high-order goods such as furniture and televisions. They require a large threshold population, have a large range and are therefore located only in accessible locations within reach of many people and where profit can be maximised (see Figure 3.6).

Figure 3.6 **Department store in a regional centre**

D. Barker

At the lower end of the hierarchy are the smallest central places, such as villages. These are much closer together, are greater in number, have only a few shops and services, have a relatively small catchment area and have a low centrality. The village shop at Hacheston, Suffolk, is profitable even in this small settlement of only 332 people (2001), because the goods on sale have low range and are bought frequently, enabling the threshold population to be met. In order to survive, many village shops incorporate a range of functions (see Figure 3.7).

D. Barker

| Figure 3.7 | A typical village shop |

Central place theory: the marketing principle

Walter Christaller attempted to describe and explain the distribution of central places by creating a model known as the **central place theory**. This was based on a study of settlements in southwest Germany in the 1930s.

Figure 3.8 and Table 3.1 show the arrangement of central places in terms of size, spacing and catchment areas that would develop in theory on a uniform (isotropic) surface.

Assumptions made in devising this model include the following:
- The region is a plain with an even distribution of population and resources.
- There is equal accessibility in all areas.
- Catchment areas do not overlap (which partly accounts for their hexagonal shape).
- Shoppers travel to the nearest central place for their needs.
- Transport costs vary in direct proportion to distance.
- Entrepreneurs seek to maximise profits.

| Figure 3.8 | Hierarchy of central places and their catchment areas according to central place theory |

Catchment area

Small market town — 2nd order central place

Town — 3rd order central place

Village — 1st order central place

As a result of his research, Christaller proposed in the model that the smallest central places (lowest order) would be 7 km apart. Those in the next order would be 12 km apart, serving three times the area and population, and so on, in an arrangement called a $k = 3$ hierarchy (see Table 3.1). This is Christaller's marketing principle. The entire population in a rural area would be able to meet its needs for both high- and low-order goods at minimum cost in central places of different status, as shown in Figure 3.8. In this instance, the number of central places at successively lower levels in the hierarchy increases in geometric progression by a factor of 3 (1, 3, 9, 27, 81...).

Table 3.1 The size and spacing of central places in southwest Germany according to Christaller

Settlement classification in a rural area	German term after Christaller	Distance apart (km)	Settlement population	Catchment area (km²)	Catchment population
Village	Markort	7	800	45	2700
Small market town	Amstort	12	1500	135	8100
Town	Kreisstadt	21	3500	400	24 000
Sub-regional centre	Bezirksstadt	36	9000	1200	75 000
Regional centre/ county town	Gaustadt	62	27 000	3600	225 000
Regional capital	Provinzhauptstadt	108	90 000	10 800	675 000

Evaluation of Christaller's marketing principle

Value of the model

Despite its limitations, central place theory does have value, both from the academic and the practical viewpoint:

- Like most models in geography, it provides a useful basis for comparison with the real world.
- The theory emphasises the mutual dependence between service centres and their catchment areas.
- It demonstrates the importance of the fundamental principles of threshold and range.
- The idea of a functional hierarchy is a significant element in explaining the size and spacing of central places and service provision in a rural area.
- The principles on which the theory is based are of value to the town and country planners of district councils. The locations of key settlements, selected services and even New Towns are important in sustaining and developing service provision in rural areas, especially where population changes have occurred.

Criticism of the model

Applying central place theory to current settlement patterns is difficult for the following reasons:

- Central places today are not necessarily just service centres; they may have other functions that influence their size and significance.
- The theory was devised at a time when transport of goods and personal mobility were more limited.
- In the 1930s, central places were the only providers of goods and services and there was stronger allegiance of customers to the settlements closest to them.
- Today, with greater car ownership, reduced travelling times, the development of out-of-town superstores, internet shopping and migration of population, there is less reliance on original, local service centres.
- New and improved roads, which affect travelling times, distort the shape of catchment areas; overlap is created by customers being influenced by factors other than distance.
- Some larger centres have become significantly more dominant in the settlement hierarchy at the expense of service provision in other less important centres nearby.
- The assumptions of the isotropic surface rarely apply in the real world in terms of uniformity of relief, population distribution, tastes and wealth.

Case study: the upper Conwy basin, North Wales

The upper Conwy basin is a well-defined geographical unit. It is an area of north Wales drained by the River Conwy and its tributaries above the small market town of Llanrwst. The rural settlements here are linked to the urban settlements further down the valley below Llanrwst; these include the towns of Llandudno and Conwy.

This area under consideration is part of a glaciated upland, which lies within the Snowdonia National Park. Mountain peaks over 500 m dominate the landscape, especially to the west, the highest being Moel Siabod (872m). Hill sheep farming, forestry and tourism are the main economic activities.

There is a dispersed pattern of isolated farmhouses on the lower mountain slopes; the main roads and nucleated settlements are confined to the valley floors. Figures 3.9 and 3.10 show the settlements and services in this area in 2005. They form a hierarchical system of central places, which provide a wide variety of high- and low-order goods to the resident population and the many visitors.

Figure
3.9

The location of shops and services in the upper Conwy basin

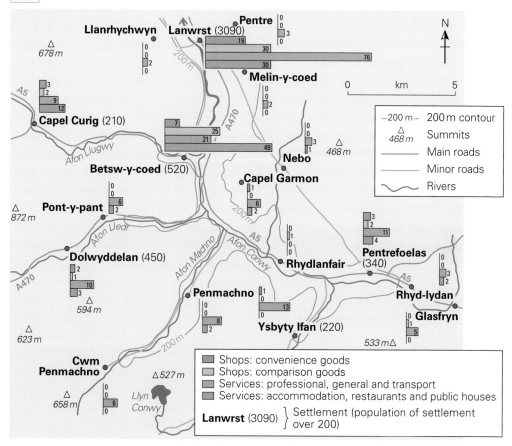

Figure
3.10

The relationship between population size and service provision in the central places of the upper Conwy basin

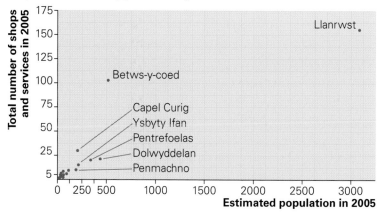

The traditional service hierarchy that has developed in the upper Conwy basin has been subject to change, especially in the last 30–40 years. This change has occurred as a result of the following factors:

- rural depopulation and loss of population from the smaller centres such as Penmachno
- growth in tourist activity, for example near Betws-y-coed
- upgrading of through routes such as the A5 and the A470
- development of large superstores in Llanrwst and the trend for weekly shopping
- the growth in ownership of second homes and holiday cottages
- the increase in personal mobility and subsidy of public transport
- planning initiatives; Menter Nant Conwy was formed to implement local regeneration.

Activity 5

(a) With reference to Figure 3.9, describe the characteristics of the settlement hierarchy and the pattern of service provision in the upper Conwy basin.

(b) With the aid of information provided in Figure 3.10, suggest reasons for the dominance of Llanrwst as a service centre.

(c) Suggest reasons why Betws-y-coed has such a high number of shops and services even though it is only 5 km from Llanrwst. It may be helpful to access www.betws-y-coed.co.uk and www.snowdonia-npa.gov.uk.

(d) The population of Penmachno was at its peak of 1755 in 1881; today it has fewer than 200 permanent residents. Referring to Figures 3.9, 3.10 and www.visionofbritain.org.uk/index.jsp, suggest reasons for its decline.

(e) How and why does the settlement hierarchy in the upper Conwy valley differ from that predicted by the Christaller model?

(f) With reference to Figure 3.10, describe the relationship between population size and service provision in the central places of the upper Conwy basin. How would calculation of Spearman's rank correlation coefficient help you to assess this relationship?

4 The impact of rural depopulation and rural–urban migration

This chapter considers the causes and the effects of **rural depopulation** and **rural–urban migration** on rural settlements and their inhabitants. As a result of these processes, the status or centrality of small market towns and villages within a rural settlement hierarchy may change over a relatively short period of time. The changes may affect the entire **social** and **economic sustainability** of some rural settlements.

Where change has been particularly dramatic, the future economic health of some rural settlements may be at risk. This chapter therefore also considers the **planning responses**. Their effectiveness is evaluated in Chapter 5, with reference to case studies at local, regional and national scales.

Overview

Rural depopulation is the absolute decline of population living in a rural area. **Rural–urban migration** is the permanent change of residence of an individual, family or group of people from an area classified as rural to one that is urban.

Many different types of rural area have been adversely affected by these two processes. In MEDCs these include:
- upland regions such as Snowdonia, the Lake District and the Massif Central in south central France
- isolated peripheral regions, for example peninsulas in Europe such as southern Italy and southwest England
- areas affected by political change, such as east European countries
- remote continental interiors, such as central USA

In LEDCs there is a tendency for most rural areas to supply migrants to the urban areas (Figure 4.1). These include:
- areas that are increasingly marginal for farming, such as parts of the Grijalva Basin, Chiapas, Mexico and Kano Province, northern Nigeria

- areas of increasing landlessness, perhaps as a result of subdivision of holdings on inheritance, such as in the Nile Delta, Egypt
- areas where commercial plantation agriculture has been introduced, such as northeast Brazil (sugar cane) and the southwest of the Malay Peninsula (rubber)
- areas of increasingly high population density where there is overpopulation, such as the deltaic lands of Barisal in southern Bangladesh

Figure 4.1 Egypt: internal migrant flows to Cairo, 1996–2004

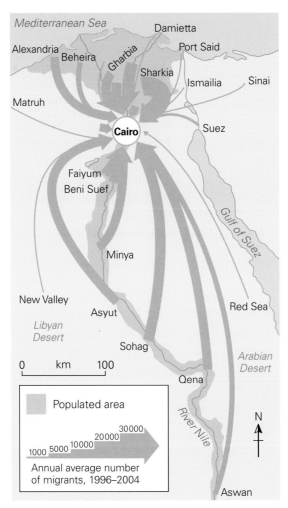

Rural depopulation in MEDCs has occurred on many occasions in the past. For example, for the northern Pennines, rural depopulation was a feature of the late nineteenth century; for north Wales, it was significant in the interwar period and the Depression of the 1920s and 1930s; in East Anglia, it was a post-Second

World War phenomenon of the 1950s and 1960s. Some areas have been affected by population loss in all of these time periods, with continuation of the process up to the present time (see Figure 4.2).

Figure 4.2   **Population change in the rural settlements of Sedbergh, Yoxford, Stogumber and Penmachno, 1801–2001**

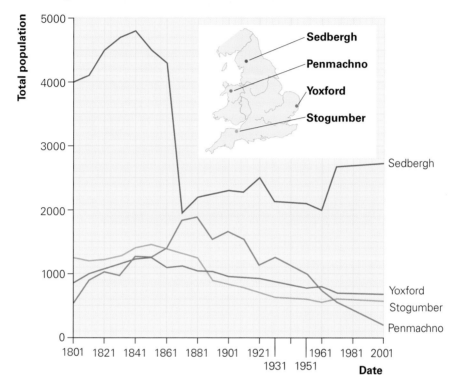

Activity 1

With specific reference to Figure 4.2, compare and contrast the population changes in the four rural settlements.

The term **urbanisation** is used to describe the increase in the proportion of people living in towns and cities in a country. **Urban growth**, on the other hand, refers to an absolute increase in the numbers of urban dwellers. Both phenomena are the result of large-scale rural–urban migration and high rates of natural increase in the urban areas. Rural–urban migration is the main process responsible for rural depopulation in LEDCs. In many LEDCs, especially in Africa, southeast Asia and central America, rates of urbanisation have been high in the late twentieth and early twenty-first centuries (see Figure 4.3).

| Figure 4.3 | The percentages of urban population for selected countries in 1991 and 2004 |

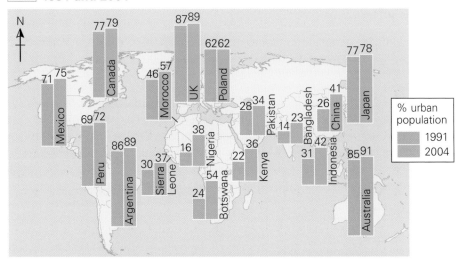

Activity 2

(a) With reference to Figure 4.3, describe the main changes in the percentage of urban populations between 1991 and 2004.

(b) What is meant by each of the following terms?
- rural depopulation
- rural–urban migration
- urbanisation
- urban growth

Causes of rural depopulation

The basic demographic reasons for rural population decline are net migration loss (out-migration exceeds in-migration) and natural decrease (death rate exceeds birth rate). The specific reasons for rural depopulation are many and diverse, depending on the locality and time period involved.

Rural depopulation in Europe in the nineteenth century

The decline of the rural population in nineteenth-century Europe was mainly due to economic rural–urban migration. Rural populations migrated to:
- coalfield areas, such as Lancashire, the central valley of Scotland and the Ruhr in Germany, which provided rural migrants with the prospect of regular and higher wages

- manufacturing towns, ports and other rapidly growing industrial centres that offered employment and a wider range of opportunities than could be found in the rural areas

Even though housing conditions were poor and the towns and cities were unhealthy environments in which to live, these economic urban pull factors were dominant.

One other cause of rural depopulation was so significant during this period that it deserves particular mention. Potato blight is a fungal disease that devastated the main food crop and staple diet of the people of southern Ireland in 1845–46, causing widespread famine. In this instance, both high mortality and out-migration led to rural depopulation. Approximately 750 000 people died of starvation and many thousands left the land to seek employment in towns and cities in the UK and USA.

The potato famine was not the only cause of rural depopulation in the south of Ireland in the mid-nineteenth century, however. Only a single heir could inherit a farmer's land, so emigration became the only alternative for the offspring who would not inherit; there were few industrial centres in Ireland providing alternative employment, and there were many attractive opportunities in the UK and the 'New World'.

So great was the impact of these factors that net increase in population in rural southern Ireland was not recorded for a century, until after the Second World War.

Rural depopulation in the UK in the twentieth century

In the first half of the twentieth century, rural depopulation continued to affect many peripheral regions. For example, in the UK, the main causes were the harsh way of life and lack of social opportunities available in remote islands, such as the Isle of Arran, and in upland regions, such as the Lake District. In the 1930s, when farmers needed to reduce costs in response to agricultural depression, much land under the plough was changed to grass; consequently, the demand for farm labour fell. During this period, too, there was increasing awareness of the attractions of urban areas, thanks to the development of rural bus services and free places in urban grammar schools becoming available to pupils from rural areas. The supply of rural housing stock was also limited, and many services available in urban areas, such as electricity, contributed to the desire to leave the countryside in the 1930s. Periodic enlistment into the armed forces and the consequences of war in the twentieth century was another factor.

It was mainly the young who migrated from rural to urban locations, leaving an ageing population behind in the rural areas. As a consequence, in the areas most affected, rural birth rates were falling and rural death rates were increasing, leading to natural decrease and further decline in the population.

Since the Second World War, a number of additional factors have contributed to rural depopulation in the UK:

- the continued loss of farm employment as a result of mechanisation
- limited employment opportunities in other sectors of the rural economy, such as quarrying and forestry
- the downward spiral of service provision in rural areas, since threshold populations could not be met (ultimately, the ensuing rural deprivation and poverty have led to further out-migration)
- the closure of coal mines (many of the pit villages in South Yorkshire, such as Kiveton Park, were small settlements in rural areas)
- planning decisions to change rural land use, such as the building of reservoirs, for example Llyn Celyn near Bala, which caused displacement of rural dwellers in central Wales
- the combination of poor communications, lack of accessibility to services, relative poverty and harsh environmental conditions, especially evident in some upland areas

Activity 3

| Figure 4.4 | An abandoned Scottish crofter's dwelling on the northeast coast of the Isle of Arran |

D. Barker

Activity 3 (continued)

Figure 4.5

Location of the croft shown in Figure 4.4

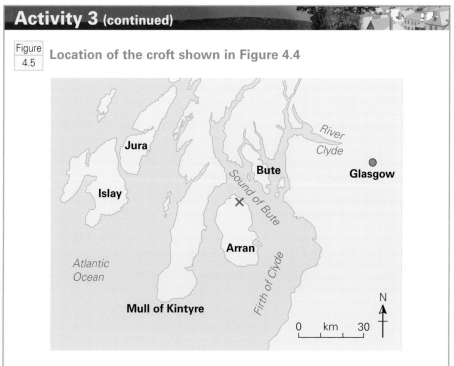

Study Figure 4.4, which shows an abandoned croft on the Isle of Arran, and Figure 4.5, which shows its location. With reference to this evidence, state and explain two possible causes of rural depopulation in MEDCs.

Push and pull factors in LEDCs

In developing countries, the decision to move to an urban area may be influenced by a wide range of factors. Essentially, these factors fall into two types: push and pull. Push factors operate in the rural areas; they are negative factors forcing migrants to leave. Pull factors are the positive attractions of the city for the rural migrant (real or perceived). In MEDCs in the nineteenth century, the strongest set of factors were 'urban pull'; in LEDCs, current flows tend to be more the result of 'rural push'.

The most significant influences on rural–urban migration are economic. For example, the employment opportunities of the cities provide a means of relieving the poverty of rural areas. Conversely, employment opportunities in rural areas may be limited; they are often reduced by the onset of commercial agricultural concerns or by competition for resources caused by rapid growth of the rural population. Systems of land inheritance, leading to landlessness, may also be a push factor, as may lack of access to water, medical care and educational facilities.

As communications improve, awareness of the potential benefits of living in urban areas (such as better quality of services) becomes a motivating force. Family members already established in the city may provide a base and contacts for others to follow. Enhanced prospects of finding a spouse is another possible reason for heading to the city. Traditionally, young men were the most likely to leave the countryside, but in recent years many women have also taken the opportunity to seek employment (although this varies between region and continent).

In LEDCs, the impact of environmental hazards also results in the loss of population from the countryside. In the short term this could be the result of an event such as a mudflow, a volcanic eruption, an earthquake, a hurricane or flooding. In the longer term, gradual degradation of the environment, such as soil exhaustion or the effects of drought, could lead to significant loss of population.

One extreme cause of rural depopulation in Zimbabwe is illustrated by the newspaper article in Figure 4.6. This is the direct result of political influences under a form of dictatorship.

Figure 4.6 Newspaper article — *The Times*, 24 October 2006

More whites to lose farms

MORE white farmers were handed eviction notices in Zimbabwe yesterday as the Government vowed to forge ahead with its land reform policy.

The seizure of about 4,000 farms since 2000 is seen as a main cause of the economic crisis in the country, but the farmers' union said that about 10 per cent of the 500 remaining farmers had now been told to leave. 'So far 40 eviction notices have been given out', Emily Cookes, the spokeswoman for the Commercial Farmers' Union, said. Those affected were from the eastern Manicaland and south-eastern Chiredzi district. The owners of several timber and coffee plantations were among those told to go. The union said that the latest evictions also threatened the livelihoods of about 3,000 farm workers.

© AFP (2006)

Consequences of rural depopulation

The effects of depopulation on rural settlements and their inhabitants are varied. They can be classified as economic, social, political, environmental and demographic. Not surprisingly, these effects may impact directly on the economic and social sustainability of many rural settlements and rural settlement hierarchies. Rural poverty, deprivation and social exclusion may ensue. The following examples illustrate the complexity of the issues that can result from rural depopulation.

Activity 4

(a) Define and distinguish between economic, social, political and demographic factors that influence rural–urban migration.

(b) Draw up a table with column headings as shown:

Economic factors	Social factors	Political factors	Environmental change	Demographic factors

Attempt to identify specific reasons for rural depopulation in each category according to this classification.

Ageing population

Not only does rural depopulation lead to a decline in the population of rural settlements and rural regions, it also causes a change in age structure. Out-migration is age-selective; the young, economically active and reproductive age groups are the first to move away, leaving an ageing population.

Övertorneå is a small town in the far north of Sweden, located on the Sweden–Finland border, 990 km north of Stockholm and 15 km south of the Arctic Circle. It lies in the Torne Valley, a sparsely populated region with a density of just 1.4 people per square kilometre.

Between 1960 and 1980, Övertorneå lost 20% of its inhabitants, with the population falling from 6250 to 5000; the projected population for 2025 is 3500.

Reasons for out-migration to urban areas further south include:
- the amalgamation of small farms into larger units
- seasonal unemployment in agriculture and forestry
- remoteness — this peripheral area of northern Sweden has few attractions for the young

Övertorneå's current average age is 45 years and this is expected to rise as the population continues to decline. The main reason for this ageing is that there are relatively few women of fertile age. As in other areas of Europe that have experienced decades of out-migration of young adults, Övertorneå's age structure does not favour population growth by natural increase. In fact, Övertorneå has reached a stage of advanced ageing where further decline and an increase in the dependency ratio seem unavoidable (see Figure 4.7).

Figure 4.7 **The age–sex population structure of Övertorneå, 2003**

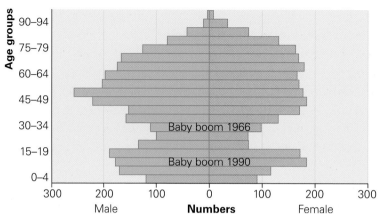

Sweden's current regional policy aims to promote development in the area. However, government schemes (such as the offer of free housing and free child-care) have failed to stem the net migrational loss.

However, it is interesting to note that the work of the Swedish economist and planner Lahti has improved the prospects of Övertorneå. It has become Sweden's first eco-municipality. Following previous loss of services, it is becoming more sustainable within the settlement hierarchy of the Torne Valley. This has been possible with the residents agreeing to use renewable energy, greater reliance on public transport, organic farming and preservation of rural land. As a policy for sustaining rural settlements, this is now supported by the Swedish government.

Many rural counties in the midwest region of the USA are interested in arresting their ageing populations and consequent rural decline by adopting a similar approach.

The change in attitude among the residents and the communal approach to development at this scale is similar to that successfully introduced by Hazare in Maharashtra (see Chapter 5). This too has received national government recognition.

Activity 5

With specific reference to Figure 4.7:
(a) describe the age–sex structure of Övertorneå
(b) suggest reasons for the age and sex patterns shown
(c) state and explain two possible consequences for Övertorneå as a result of its age–sex structure

Loss of shops and services

Depopulation affects service provision in many rural settlements. The spiral of decline, or negative multiplier, can be explained by continued loss of customers. Threshold populations of shops and services cannot be met, leading to closure; this becomes a reason for further loss of population, and so on. At first it is the high- or middle-order services (e.g. chemists, hardware) that are lost, since they require higher thresholds to be economically viable; lower-order services (e.g. butchers, bakers) are then lost, until few services remain. Any surviving shops must diversify, with an emphasis on the variety of services and retailing they offer (e.g. general store, sub-post office).

Depopulation and service decline are not just features of remote upland regions such as the Massif Central in France, or declining peripheral regions such as the Mezzogiorno in Italy. Depopulation has also occurred in densely populated lowland areas, such as central Suffolk.

Case study: shop and service decline

Earl Soham is a village in Suffolk where population decline has occurred over a long period of time (see Figure 4.8).

| Figure 4.8 | **Population decline in Earl Soham, 1861–2001**

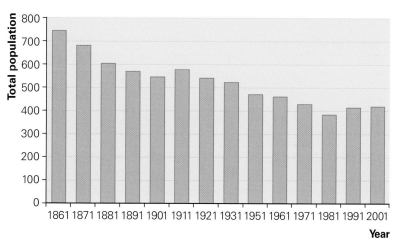

The village has lost most of its shops and services; consequently, its status within the settlement hierarchy of north Suffolk has declined significantly. Suffolk County Council has adopted a **key settlement policy** to arrest this decline. Shops and services have been concentrated in nearby Framlingham to ensure that essential services are available to the population in this area.

Figure 4.9 Photographic evidence of former shops and services in the village of Earl Soham in Suffolk

Grocer

Newsagent and stationer

Butcher

The smithy

Post office

The Falcon Inn

All images D. Barker

Figure 4.9 shows the evidence of service decline, with many buildings formerly used for shops and services converted into private residences. The large shop windows and the house names adopted by the residents, such as 'The Old Stores', point to the wider range of services available in the past. Since the 1960s, the village has lost its baker, butcher, grocer, post office, newsagent/confectioner, garage and a public house. Today just two shops remain, providing a mixture of basic convenience goods and some specialist items. The dilemma for the villagers is this: although the population of Earl Soham has grown slightly in the last 20 years, service provision has continued to decline.

However, depopulation is not the only reason for service decline. Other factors include Earl Soham's accessible location on the A1120 (to Stowmarket and Ipswich); the increase in personal mobility as a result of greater car ownership and more frequent bus services, which give villagers easy access to goods at cheaper prices in the supermarkets of the market towns; second home owners often purchasing essential items elsewhere before their visit to the village; bulk buying/weekly shopping at out-of-town stores (a trend arising from the shopping behaviour of the many commuters who now live in the village while retaining their urban connections).

It is a relatively simple task to record data in the field, by personal observation, on a large-scale map such as the 25 inch to 1 mile map of Earl Soham for 1904 (Figure 4.10); this could include age and function of buildings. Housing and other land uses can be added or deleted to show physical change in the village since this date.

Figure 4.10 **Part of the OS 25 inch to 1 mile map of Earl Soham, 1904**

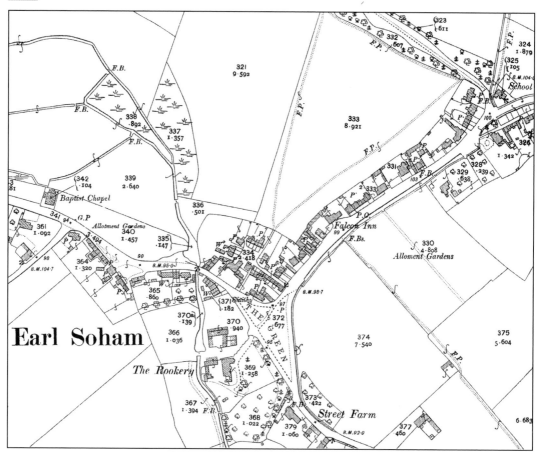

In this instance, the smithy (Forge House), the Falcon Inn and the post office (see Figure 4.9) are now private residences; the school has expanded and the allotment gardens have become a bowls green and tennis court.

Activity 6

Conduct a field investigation in a rural settlement where there has been population change. Use a large-scale OS map (available in County Records Offices for a range of dates) to record the past and present functions of each building and open space.

The problems of service decline at Earl Soham are common to many Suffolk villages. Surveys conducted by Suffolk ACRE (a charitable organisation) on behalf of the Rural Development Commission have shown that service decline in the county has been worse than for England as a whole (see Table 4.1).

Table 4.1 Percentage of parishes lacking key services in Suffolk and in England

Service	Suffolk	England
Permanent shop	50	42
General store	79	70
Post office	48	43
Public house	33	29
Daily bus service	80	75
School (for any age)	62	49
School (for 6 year olds)	64	50

Activity 7

Answer the following essay question. Be sure to refer to named rural areas and specific groups of rural dwellers.

'Explain how changes in service provision affect the lives of some rural dwellers more than others.'

In summary, the effects of rural depopulation on service provision in rural settlements are shown in Figure 4.11.

Figure 4.11 **Effects of rural depopulation**

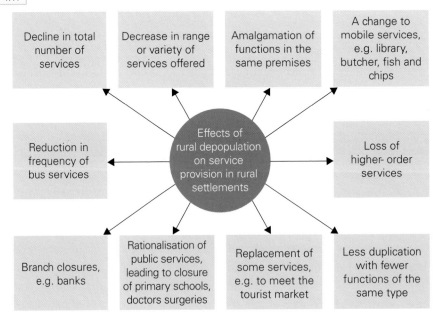

The issue of rural deprivation: cause or consequence of rural depopulation?

Rural depopulation is responsible for deprivation in some rural areas, but not all. Out-migration is one of many factors accounting for rural deprivation and social exclusion. In fact, in many instances it is a symptomatic response to environmental problems or socioeconomic difficulties, rather than being an underlying cause.

In Scotland, for example, poverty and social exclusion are widespread in rural areas that have experienced population decline. In the northwest Highlands and Islands, this has led to loss of jobs and loss of services; many rural settlements have become unsustainable and rural poverty is increasing. These problems are linked to low population densities following long periods of rural depopulation. The geography of this region, which has many island communities, numerous sea lochs and incised glaciated valleys, makes communications difficult and this has contributed to the problem. Details of social and economic indicators can be found in Scotland's 2001Census at: www.gro-scotland.gov.uk/census/

However, in west Cornwall, rural poverty is more the result of poor access to services, unaffordable housing stock (partly due to second home ownership and a low proportion of social housing to rent), high levels of unemployment (often because of the seasonality of the work) and low wages, rather than the effects of rural depopulation.

Figure 4.12 shows some indices used to measure rural poverty.

| Figure 4.12 | **Indices of rural poverty** |

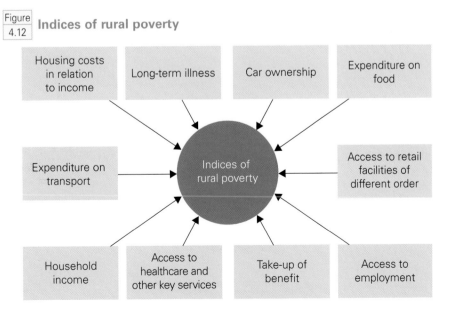

The issue of second home ownership

There has been growth in second home ownership, including holiday cottages, in many European countries. At a national level, this growth tends to be the result of a buoyant economy, strong growth in house prices, and the ready availability of finance. In terms of individual/family decisions, it is often encouraged by the prospect of rising house prices. Furthermore, changes in work patterns enable people to work from a rural second home some days and at the office from their main urban-based home on other days. This trend is due partly to the desire for an improved quality of life and it is made possible by rising car ownership and increased personal mobility.

The main reason for having a second home is to have a weekend cottage or retirement home, which at the same time is a safe investment. In the UK this trend is likely to continue as the proportion of older affluent people increases. In addition, there has been a growth of younger cohabiting couples with two homes. The advantage is that unmarried couples, for tax purposes, can divide their two homes, making a second home a tax-free investment.

The spiral of rural decline

Second home ownership is linked to the spiral of decline in rural areas (Figure 4.13). It leads to economic and social unsustainability of rural settlements, contributes to ageing population, loss of services, deprivation and social exclusion, and ultimately results in rural depopulation.

Figure 4.13 **The spiral of rural decline**

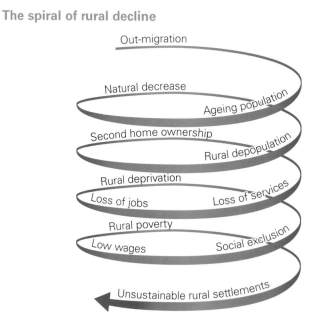

Second homes are in such high demand in attractive or scenic areas that property prices are driven up and put out of reach of local residents, especially the young. Unless rural district councils incorporate lower-cost starter homes for rural dwellers on relatively low incomes, there is further reason for them to migrate to an urban area, leaving the more elderly population remaining in the village.

Case study: the effects of second home ownership

Lozère is a *départment* in southern France, located in the upland region known as the Massif Central. It is a rural area with a long history of depopulation. Its total population fell by 9% from 81990 in 1962 to 73753 in 2004. During this period, even though net migration has changed from severe loss to recent gain, continued natural decrease has meant that population growth since 1990 has been limited (Figure 4.14).

Typical examples of small rural settlements in Lozère that have experienced population decline and growth in second home ownership are Laubert and Arzenc-d'Apcher. Between 1999 and 2005, Laubert's population fell from 58 to 46 while the number of second homes increased from 16 to 21. For Arzenc-d'Apcher the population fell from 133 to 111 and the number of second homes increased from 46 to 60. Although these are small numbers, over an entire region the social and economic effects are highly significant. Key data in English for all *communes* is now available in the French census at: www.insee.fr/en/recensement/page_accueil_rp.htm

Figure 4.14 **Average annual rates of net migration and natural increase/decrease, Lozère, 1962–2004**

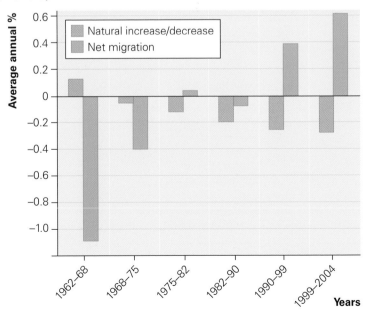

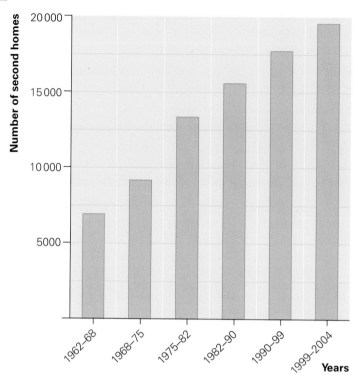

Figure 4.15 Second home ownership in Lozère, 1962–2004

Number of second homes (y-axis)

Years (x-axis): 1962–68, 1968–75, 1975–82, 1982–90, 1990–99, 1999–2004

Activity 8

(a) With reference to Figure 4.14, describe the changes in net migration and natural increase/decrease in Lozère between 1962 and 2004.

(b) With reference to Figure 4.15, describe the changes in the number of second homes in Lozère between 1962 and 2004.

(c) State and explain the possible effects of increasing second home ownership on:

(i) shops and services

(ii) the built environment

(iii) the values and attitudes of the permanent residents of small settlements in rural areas, such as the Massif Central

A number of villages in Lozère with a high proportion of second homes and a population of less than 1000 have no shop at all. The remaining, mainly elderly residents are supplied by weekly grocers' rounds plus other *'services aux particuliers'*, i.e. mobile services provided to individuals in their home or village, such as hairdressers and butchers.

Activity 9

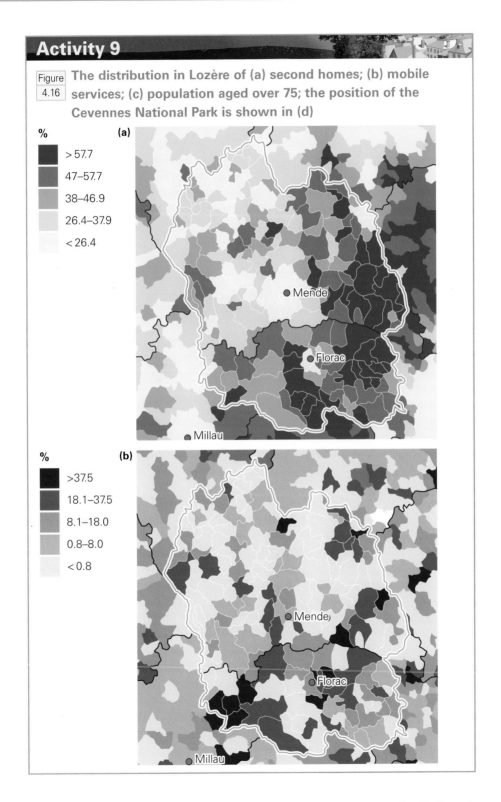

Figure 4.16 The distribution in Lozère of (a) second homes; (b) mobile services; (c) population aged over 75; the position of the Cevennes National Park is shown in (d)

(a)

%
- > 57.7
- 47–57.7
- 38–46.9
- 26.4–37.9
- < 26.4

(b)

%
- > 37.5
- 18.1–37.5
- 8.1–18.0
- 0.8–8.0
- < 0.8

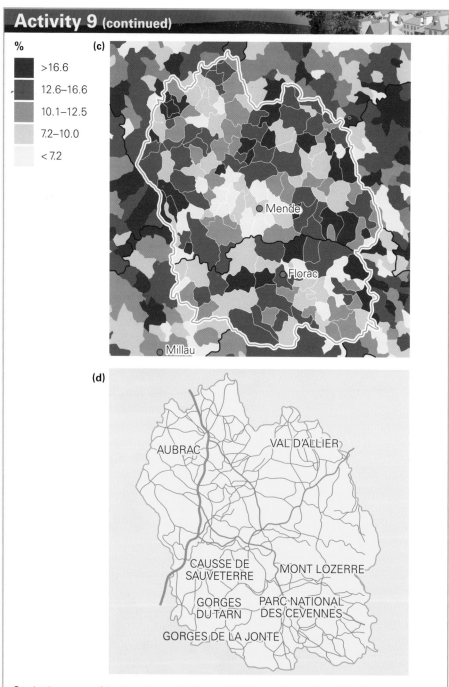

%

■	>16.6
■	12.6–16.6
■	10.1–12.5
■	7.2–10.0
■	< 7.2

(c)

Mende

Florac

Millau

(d)

AUBRAC

VAL D'ALLIER

CAUSSE DE SAUVETERRE

MONT LOZERRE

GORGES DU TARN

PARC NATIONAL DES CEVENNES

GORGES DE LA JONTE

Study the maps of Lozère shown in Figure 4.16. Describe and explain the similarities and differences that appear to exist in the locations of second homes, mobile services and the elderly population.

While mobile services might alleviate rural deprivation in small remote upland villages and hamlets in southeast Lozère, a further attempt has been made to sustain many of the larger villages that act as service centres for a wider area in rural France. Under the so-called '1000 French villages' project, a multi-service store can be set up on application for aid and financial support from the local council. Often this is established by the conversion of the last remaining shop into this new upgraded format. The purpose of this project is to revitalise French villages with a population under 2000 by maintaining and restoring businesses, providing basic public services at focal points and by reviving social contact and cultural activity.

This is seen to be important where second home ownership is high because it leads to strained relationships between the resident population and the weekend and holiday visitors. Often, land and buildings are acquired which might otherwise have been used in the process of agricultural improvement. Conflict also arises in terms of traffic, parking, rising property values and lack of consideration for those who live and work in a 'holiday area'. These issues are particularly problematic in the Cevennes area of southeast Lozère.

5 Case studies: rural depopulation and the planning responses

The causes and effects of rural depopulation are discussed in Chapter 4. Their considerable diversity is best illustrated in a series of case studies at different scales.

This chapter considers rural depopulation and the planning responses at the national (USA), regional (English Lake District) and local (Ralegan Siddhi) scales. Each case study has a threefold structure:

- demographic change
- socioeconomic problems
- evaluation of the planning/management responses

It should be understood at the outset that while the title of this chapter concerns rural depopulation, this is merely the net effect of population change in the three chosen case studies. Concealed within this net loss are more complex population changes, including urban–rural migrant flows, international migration, and changes in age and sex structure. The ensuing problems therefore require careful planning and, because of increasingly rapid change in some rural areas, these management plans are in need of constant review. The contrast in approach to planning between areas in MEDCs and LEDCs is also a significant feature of this chapter.

USA

Demographic change

A significant proportion of counties in the USA have experienced rural depopulation since 1970. In fact, the rate of depopulation in 232 counties has accelerated, especially since 1990 (see Figure 5.1).

Activity 1

With reference to Figure 5.1 and an atlas map of the USA, describe the pattern of rural depopulation between 1970 and 2000.

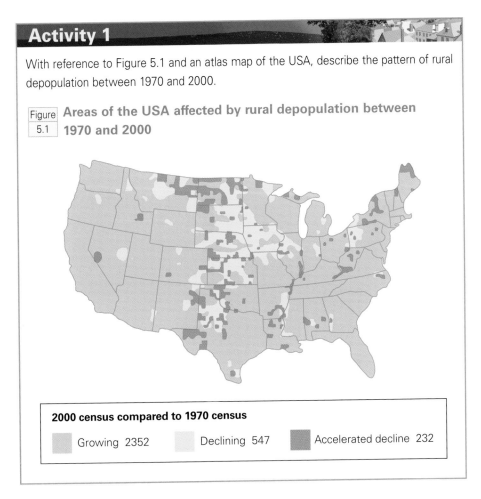

Figure 5.1 **Areas of the USA affected by rural depopulation between 1970 and 2000**

2000 census compared to 1970 census

Growing 2352   Declining 547   Accelerated decline 232

The two largest geographical areas suffering depopulation are the Great Plains, traditionally an area of wheat production and livestock ranching, and the Midwest, often referred to as the Corn Belt. The counties that have experienced population decline are all economically dependent on agriculture.

The major cause of depopulation is out-migration. The main reasons why farmers and their families have migrated to urban areas is the dramatic advance in technology. Farms are managed and operated with far fewer workers than even 10 years ago.

Service decline is also contributing to the population loss in many smaller towns and communities. These are entangled in the downward spiral of service provision as thresholds are undermined. Smaller rural settlements cannot compete with urban areas, which offer a far greater number and variety of services, and many are becoming economically and socially unsustainable.

The demographic effect is illustrated by the contrasting population pyramids for Linn County and Worth County in Figure 5.2. The locations of these counties are shown in Figure 5.3.

Figure 5.2 **The contrasting population structures of Linn County (metropolitan) and Worth County (rural) in Iowa, USA 2000**

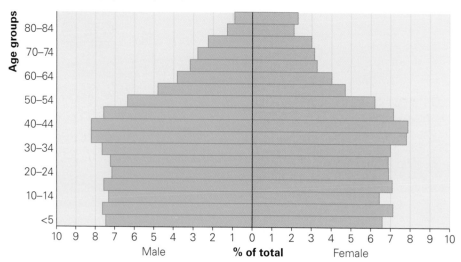

Linn County, Iowa
Total population   191 701
Density (population per square mile)   267.2

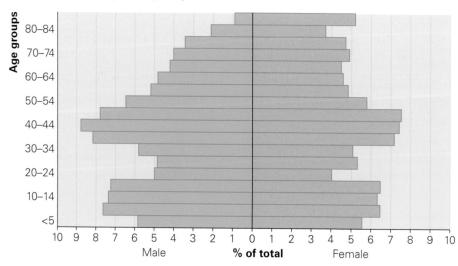

Worth County, Iowa
Total population   7 909
Density (population per square mile)   19.8

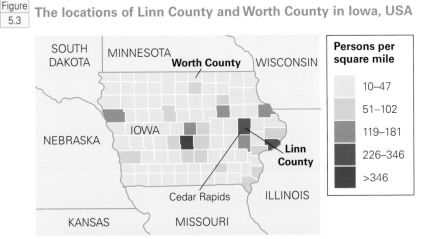

Figure 5.3 The locations of Linn County and Worth County in Iowa, USA

Activity 2

Study Figures 5.2 and 5.3. Compare the population structure of Worth County with that of Linn County. Suggest possible reasons for the differences in age and sex structure.

Socioeconomic problems

In Worth County, as in many other similar Midwest rural counties, one consequence of the high proportion of elderly and few young adults is natural decrease. This, together with out-migration, adds to rural population decline.

These lower populations support fewer retail establishments and agricultural businesses, such as suppliers of farm machinery, chemicals and feed. Increasingly, such businesses are found only in larger urban centres, such as Cedar Rapids (see Figure 5.3). Moreover, depopulated communities cannot support vital infrastructure. This includes government agencies (fire and police), roads, schools and hospitals. As the population decreases, the cost per head of supplying these services increases, creating a poor social and economic environment, and so the spiral of decline continues.

Approximately 50% of rural banks in Iowa have closed since 1984 and the customer base is likely to shrink even more in areas of ageing population and increasing death rates.

Planning/management responses

The US economic development strategy is to use government funds to reverse market forces and to restore economic and social vitality to declining rural areas. So far the lack of success of this policy is mainly due to its scale. Geographically, the problems cover a vast area, the largest part of which — the Great Plains — stretches in a 200 km belt east of the Rockies for over 2000 km from the Canadian border to the Gulf of Mexico. The rate of population decline has been particularly rapid in the last decade. The government at federal and state levels has not been able to respond effectively to rural decline, partly because it is a fairly recent problem and partly because it has happened so quickly.

More successful, further south in Kansas, have been the efforts of several communities to give away land to attract new residences and businesses. But this local solution is only on a small scale and does not reverse the trends of the entire Great Plains. It is thought that increasing use of the internet might help businesses to market beyond their county lines. However, in the case of banking, this has merely helped larger banks in the cities to compete with the local banks more successfully. Retailing has been badly affected for the same reason.

The English Lake District

The Allerdale district of Cumbria covers the northwest quarter of the English Lake District. Its southern boundary is the high watershed that divides the glacial troughs of Buttermere/Crummock Water and Ennerdale and it extends northwards to include the coastal lowlands of the Solway Firth.

The area in Figure 5.4 has experienced rural depopulation between the last two census dates. Physiographically, this is largely an upland glaciated region drained by the rivers Derwent and Cocker. These rivers have their sources on the slopes of the Buttermere and the Borrowdale Fells, which reach up to 930 m.

Figure 5.4 **The remote rural parishes of Buttermere and Above Derwent, viewed from Red Pike**

D. Barker

This is a harsh environment of steep slopes, bare rock, thin soils and cold climate. Seathwaite village has reputedly the highest annual rainfall total in England (over 2500 mm). Accessibility is poor; the B5289 road, when not congested by tourist traffic in the summer, is often impassable in winter snow. Its steep gradients and narrow bends are difficult to negotiate for delivery lorries and other vital services. Even Keswick, the largest settlement in the area, experienced population loss between 1991 and 2001. Traditionally, the main occupations have been in pastoral farming, quarrying and forestry. Now that these are declining, tourism is the most important source of income within the National Park.

Demographic change

Where the valleys of the Derwent and Cocker converge at Cockermouth, they enter the coastal plain. This is an area of higher population density, with towns and villages. Most of this coastal lowland has experienced population gain since 1991.

Population change between 1991 and 2001 in this selected part of Allerdale is shown in Figure 5.5.

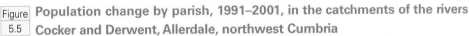

Figure 5.5 | **Population change by parish, 1991–2001, in the catchments of the rivers Cocker and Derwent, Allerdale, northwest Cumbria**

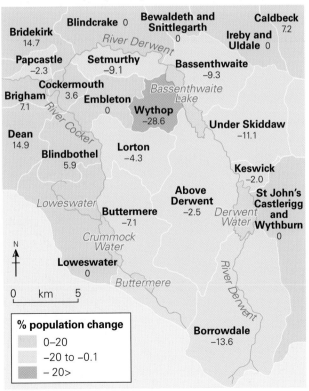

The areas of greatest loss have been in the upland fell areas, which are some of the highest and most rugged parts of the Lake District. These areas form the upper Derwent and Cocker valleys. Here, all parishes except Loweswater and St John's Castlerigg and Wythburn (no change) experienced population loss in the decade 1991 to 2001. Losses ranged from 28.6% (Wythop) to 2.0% (Keswick). Borrowdale, Buttermere, Under Skiddaw and Bassenthwaite also experienced significant depopulation.

Population growth can be seen in the parishes of the lower areas to the north and west of the area. These include the town of Cockermouth and its neighbouring parishes of Bridekirk, Dean and Brigham, all on the lower land beyond the designated boundary of the Lake District National Park.

Activity 3

Figure 5.6 Age structure of selected parishes in Allerdale, 2001

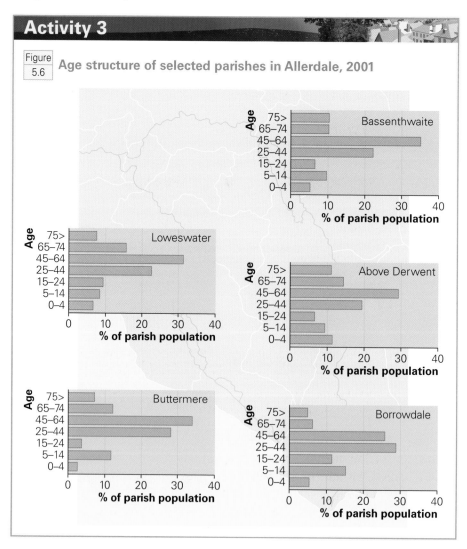

Activity 3 (continued)

(a) With reference to Figure 5.6, describe the population structure of any one parish.

(b) Suggest reasons why there are variations in the age structures shown, even within this relatively small area.

Socioeconomic problems

Service provision in most rural settlements in Allerdale has declined as the population loss has led to rationalisation. Figures 5.7 and 5.8 show the locations of GPs (health services) in 2004 and primary schools (education services) in 2005 in Allerdale and adjoining areas.

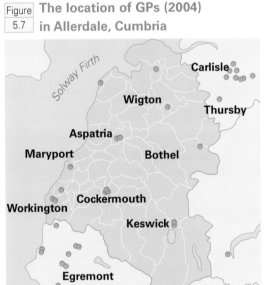

Figure 5.7 **The location of GPs (2004) in Allerdale, Cumbria**

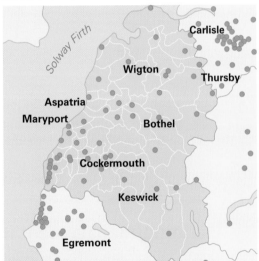

Figure 5.8 **The location of primary schools (2005) in Allerdale, Cumbria**

Activity 4

With reference to Figures 5.7 and 5.8, describe and explain the distribution of GPs and primary schools shown.

The information provided in these maps is available on the National Statistics Online website for the 2001 Population Census for England and Wales. Access 'Neighbourhood' on: www.statistics.gov.uk.

Changes in the number and distribution of post offices may be traced using current and past editions of 1:50000 and 1:25000 Ordnance Survey maps.

Based on this method, Table 5.1 shows that in the Derwent and Cocker valleys, the total number of post offices has fallen over the last 25 years.

Table 5.1 Post offices in the Derwent and Cocker valleys

Settlement	1981	1994	2006
Braithwaite	✔	✔	✔
Brigham	✔	✔	✔
Cockermouth	✔	✔	✔
Eaglesfield	✔		
Greysouthen	✔	✔	✔
Keswick	✔	✔	✔
Lorton	✔	✔	✔
Papcastle	✔	✔	
Portinscale	✔		
Rosthwaite	✔		
Uldale	✔	✔	
Total	**11**	**8**	**6**

The closure of village post offices is a considerable blow to many rural communities. Their loss from rural areas such as Allerdale causes social as well as economic problems for the inhabitants. Apart from posting and delivery of mail, post offices provide other functions, such as retailing, banking and provision of pensions, and they serve as a meeting place in the village. Closure affects some members of the community more than others, in particular the elderly and those without cars or with limited access to public transport.

Planning/management responses

In its strategic plan, the Allerdale Borough Council classifies rural Cumbria into four types of area:
- deep rural (central and southern areas of the national park)
- commuter rural (with easy access to the M6 corridor)
- ex-industrial rural (west Cumbria)
- remote rural (the Solway Plain)

It is the first of these areas, 'deep rural', that is most affected by rural depopulation and declining service provision. The uplands of Allerdale district have the following problems:
- remoteness from major settlements and connecting transport routes
- high percentage of second home ownership

- low incomes in local employment, including tourism as well as the traditional industries such as farming
- high house prices
- limited housing stock
- limited access to services

The rural settlement hierarchy in this area has changed as a result of depopulation, rationalisation and the restrictive planning policies of the National Park. In addition to population loss, an ageing population structure, limited access to vital services and rural deprivation are significant problems.

One of the main concerns is the combination of high cost of housing and low wages. The Cumbria Rural Housing Trust has shown that there is a shortage of rented housing and that reasonably priced housing is not available to buy. It is notable that some large, privately owned estates and organisations such as the National Trust and the Forestry Commission are now providing housing for local people. But there is a great need to increase the supply further, so that people can live near to work, and communities can be more balanced and not dominated by holiday and second homes.

At a different level, Cumbria County Council is responsible for service provision in the area (i.e. police, fire, health and education). At a higher level still, the North West Development Agency funds other concerned bodies, such as the Cumbria Rural Enterprise Agency, helping farmers in dispersed locations and rural businesses. In addition, SPARSE (Sparsity Partnership for Authorities delivering Rural Services), a grouping of the most rural local authorities in England, has an interest in economic development and housing; its sister group, the Rural Services Partnership, incorporates public service providers in the most rural areas.

The Commission for Rural Communities reports that Allerdale Borough Council welcomed the opportunity to increase the amount of tax raised from second home owners by reducing the council tax discount from 50% to 10%. In total this has raised £2.8 million a year from 6100 properties. There is not always agreement on how this extra revenue is best used, which is why it is important to understand the need for Local Area Committees that can input into the planning process as well as the decision makers at higher levels, and Strategic Partnerships. Specifically, Allerdale is using this money to subsidise bus fares and to support the Market Towns Initiative.

Thus a number of mechanisms and tiers of planning in Cumbria aim to reduce rural poverty and deprivation. Because of the difficult geography of the region, progress has been slow and rural services continue to decline even in villages where population decline has been reversed.

Ralegan Siddhi, India

Ralegan Siddhi is situated 140 km east of Mumbai, 80 km northeast of Pune, and 50 km southwest of Ahmednagar (see Figure 5.11 on page 98). This is a drought-prone area lying in the rain shadow of the Western Ghats. Rainfall is erratic, between 200 and 850 mm per year; temperatures range between 28°C and 44°C.

Ralegan Siddhi demonstrates the problems faced by rural communities in the developing world because of demographic change. This example has been chosen because it is also possible to evaluate the effectiveness of the planning solutions at a local scale. There has been a remarkable transformation of this village in the last 40 years, from abject poverty and despair to increased prosperity and socioeconomic sustainability.

Population change, poverty and planning in India

Figure 5.9 shows that population growth has occurred in practically every district in India between 1991 and 2001. However, the map conceals the pronounced rural–urban migration that affects many rural settlements. Rural–urban flows have been relentless, both within and between districts. Young men in particular migrate from villages to nearby towns. The flows to the megacities such as Mumbai are particularly large.

| Figure 5.9 | Growth of population in the districts of India, 1991–2001 |

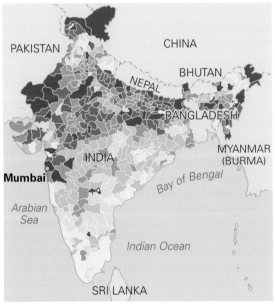

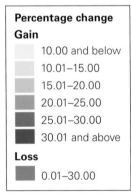

Percentage change
Gain

- 10.00 and below
- 10.01–15.00
- 15.01–20.00
- 20.01–25.00
- 25.01–30.00
- 30.01 and above

Loss

- 0.01–30.00

Activity 5

Study Figure 5.9, which shows population growth in the districts of India between 1991 and 2001.

(a) With reference to Figure 5.9 and an atlas map, describe the pattern of population change shown.

(b) Suggest reasons for this pattern.

In rural India appalling poverty, unemployment, low and uncertain wages, uneconomic land holdings and poor facilities for health, education and recreation act as push factors. The prospect of alleviating these problems pulls migrants to urban areas.

In absolute numbers there was huge urban growth in India between 1991 and 2001, from 217.6 million to 285.3 million. Yet despite the scale of the rural–urban migration, the percentage of the Indian population living in urban areas has grown only slowly, from 26% in 1991 to 29% in 2001. This is explained by continued high rates of natural increase in rural areas (Figure 5.10). The total population classified as rural in the 2001 census of population for India was 741.6 million people, or 71% of the Indian population. While urban poverty is high, especially in the slums, the absolute numbers of poor and the incidence of poverty is far greater in rural India.

| Figure 5.10 | **Changing proportions of rural and urban populations in India, 1996–2006** |

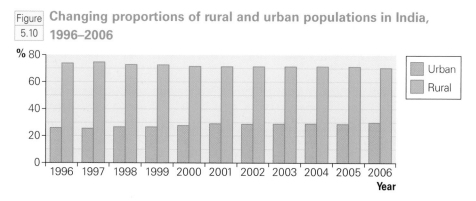

In February 2006, the Indian government launched a multi-million dollar scheme to tackle rural poverty. Work for 100 days per year is now guaranteed for every rural household. In a country where three out of every four people are rural dwellers, the ability to manage such a scheme involving almost 750 million people is questionable. Small-scale projects might be more effective in the longer term (as in Kansas and Cumbria). It has been demonstrated so far that the problems arising from rural depopulation are localised and that accurate local-scale information is an essential step in the planning process.

We saw that in Allerdale, in Cumbria, planning policies and rural planning agencies are beginning to tackle serious problems of **relative poverty** and relative deprivation. But in the case of Ralegan Siddhi, the policy makers have to deal with **absolute poverty**; by 1975, per capita annual incomes were as low as 270 rupees. In Ralegan Siddhi poverty is not just caused by demographic change but also by the effects of drought on farming and a complex set of social and religious factors. These are interrelated and not easily solved. But since 1975, local initiatives have been so effective that this settlement has become a model for planning throughout the whole of India.

Demographic change

The village population of Ralegan Siddhi stagnated in the 1960s and 1970s (1209 in 1971), followed by a period of increase in the 1990s (1982 by 1991); the projection for 2006 was 2540. The slow growth of the 1960s and 1970s was the result of out-migration and high rates of infant mortality. The more rapid growth of the 1990s onwards was the product of the village development programme started in 1975. This included female education, family planning and consequent birth control. Infant mortality rates decreased and some migrants returned from the cities.

Before the development programme, the loss of young men due to out-migration was a response to poor agricultural productivity and the pressure of local money lenders for debt repayment. The main migrant flows were to the nearby small towns of Shirur and Parner and ultimately to the larger urban areas of Ahmednagar, Pune and Mumbai (see Figure 5.11).

Figure 5.11 **Direction of rural migrant flows from Ralegan Siddhi**

Socioeconomic problems

By the early 1970s, the socioeconomic sustainability of Ralegan Siddhi was under serious threat. This was due to a variety of factors:

- age-selective migration removing many of the most energetic villagers
- lack of medical services — the nearest medical help was 14 km away in Shirur
- inadequate education facilities — there was only a two-classroom primary school
- poor transport links — there was only one bus to the nearest secondary school in Shirur
- poor levels of nutrition — one survey reported that nearly half the villagers had only one meal per day, and a third of households missed a meal every other day
- the high incidence of debilitating guinea-worm infection and other water-borne diseases
- the practice of caste discrimination, including untouchability, destroying any cooperative spirit and community feeling — for example, the harijans or untouchables were not permitted to draw water from the village well
- drunkenness and alcoholism among the men and youths of the village, causing problems of vandalism and wife beating
- high birth rates and very high rates of infant mortality
- severe droughts leading to perennial water shortages
- high levels of illiteracy, especially among women; in 1971 only 30% of the population (72 women and 290 men) were literate

Activity 6

Study the list above, which gives 11 reasons why Ralegan Siddhi was becoming increasingly unsustainable socially and economically by the early 1970s.

(a) Classify each reason as demographic, social, economic or physical.

(b) For any two of the factors in the list, explain how they contribute to socioeconomic unsustainability.

Planning/management responses

The enterprise of one man is largely responsible for the reform of this village and its inhabitants. Figure 5.12 shows Mr Anna Hazare encouraging people to participate in village development. This approach is particularly important in rural India, where communal spirit and cooperation are essential drivers in development, similar to that of self-help in the urban slums.

Figure 5.12 **Mr Anna Hazare encouraging communal participation in village development**

Prasan Firodia

Having restored the village temple using all his savings and involving a newly formed youth group, Hazare gained the confidence of the villagers. This created a meeting place from which the necessary reform could be discussed and administered.

By agreement, the most debilitating **social problems** were tackled first. This involved a complete ban on the sale and consumption of alcohol. Public humiliation by physical punishment was occasionally necessary but social pressure was usually effective enough. All illegal stills were closed by 1979. In addition, the upper castes agreed to take the initiative by removing social discrimination on the basis of caste. It was also agreed not to accept dowry at the time of a son's marriage (the practice of giving dowry has caused many problems among the poor of rural India). Gradually, the practice of bribery was removed. Superstitions such as sacrificing goats (valuable assets alive) and pretending to be possessed by the goddess Padmavati (giving power over others) were broken down.

Next, the **demographic problems** were pursued, mainly by improving the status of women. They had been suppressed; they were made to work hard at home and their husbands gave them little choice in the number of children they had. Education programmes have helped to double female literacy rates since 1971, and the status of women has improved significantly.

The implementation of strict family planning measures, with sanctions such as restricted access to food and other benefits of the development programme, helped to reduce the birth rate. This was reinforced by the Maharashtra government, which initiated an education programme plus distribution of free contraceptives. Educational facilities were developed. A secondary school was built, which has 500 on the roll and serves villages up to 20 km away.

Finally, the **economic difficulties** were overcome by the implementation of the watershed scheme. The villagers built small dams and channels along the slopes of the surrounding hills to the northeast and south. This enabled water to be stored for irrigation of agricultural land and watering livestock. The area of irrigated land increased from 55 acres in 1971 to over 700 acres today. The damming of streams allowed more water to percolate and recharge the groundwater. The water table has been raised from a depth of 20 m to 6.5 m. This has enabled easier pumping from wells for both domestic and farming uses. Employment was also created by dam construction/maintenance plus agriculture and associated tertiary activity. Ralegan Siddhi now has a bank, seven general stores, a post office, a secondary school, an animal care centre, a library, a bus station and a telephone exchange. It has developed a higher degree of centrality, becoming a catchment area for many surrounding villages. Average annual incomes have risen to more than 2200 rupees per capita.

Evaluation of the outcomes

The successful development of this village is the result of self-help and cooperative involvement at a local scale. There has been some outside financial help and planning support (stakeholders include businessmen in Pune) but, in the main, the policies have been devised and implemented by the villagers in a well-organised format with clear priorities at each stage.

The significance of this scheme lies in its example for Maharashtra and the rest of India. According to classifications used in the Census of Population for India in 2001, Maharashtra had 43 711 villages in its rural settlement hierarchy; in India as a whole there were 638 588, many of which have adopted the Ralegan Siddhi planning approach for a sustainable rural settlement.

Rural–urban migration in India in some instances can alleviate problems of overpopulation in rural areas. It can also exacerbate socioeconomic problems, since it is often the poorest who are left behind. In the case of Ralegan Siddhi, a number of earlier rural migrants have returned. Making use of their acquired skills and experience, new ideas have been introduced, new agricultural techniques have been promoted and new attitudes towards family size have helped to ease dependency burdens. Increasingly it is being recognised that there is a strong case for development programmes such as this at the local scale.

Figure
5.13 (a) Villagers preparing dams before the monsoon; (b) communal work in the fields and enhanced irrigation have increased agricultural output

(a)

Prasan Firodia

(b)

Prasan Firodia

Improvement of village conditions so that people do not wish to leave, and the alleviation of conditions for those left behind, is an effective approach to rural planning. It may provide the solution to urban as well as rural problems in the developing world.

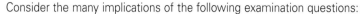

Activity 7

Consider the many implications of the following examination questions:

(a) How important is a sustainable hierarchy of rural service centres to the economic and social well-being of rural communities in MEDCs?

(b) With reference to named rural areas, describe the effects of rural depopulation and rural–urban migration, and explain the effectiveness of the planning schemes designed to sustain rural communities.

6 The impact of counterurbanisation

Many rural settlements and rural regions, especially in MEDCs, have experienced population growth since the 1960s; this is the result of counterurbanisation and urban–rural migration.

What is counterurbanisation?

Counterurbanisation is the increase in the proportion of people living in areas defined as rural, in a country or region. The two main demographic processes responsible for counterurbanisation are urban–rural migration and natural increase. **Urban–rural migration** is the permanent change of residence of an individual, family or group of people from an urban area to one that is rural. Counterurbanisation occurs as the migration causes redistribution of population down the settlement hierarchy from large urban areas to small market towns and villages. **Natural increase** is the growth in population when birth rates exceed death rates. This occurs in rural areas where the incomers, such as commuters, are predominantly young adults.

Counterurbanisation is one element of a demographic change known as **population turnaround**. This term signifies the complete reversal of long-established net migration flows from rural–urban to urban–rural. In some rural areas this has resulted in population growth for the first time in decades. It is a characteristic of the population geography of many regions in MEDCs.

Demographic turnaround has not occurred in LEDCs. In some developing countries there is evidence of the return of original rural migrants to their areas of origin after long periods in the city. But overall, numbers are insufficient to exceed the continuing huge influx of city-bound rural migrants.

This chapter identifies the characteristics and the causes of counterurbanisation. It also considers the consequences of this process for rural communities. In some instances counterurbanisation has had a serious impact on the provision of services within traditional rural settlement hierarchies and, on occasion, this has led to conflict of interest between different groups of inhabitants. These economic and social problems have prompted a wide variety of planning measures.

Chapter 7 demonstrates how these measures have been applied in two detailed case studies. This includes an evaluation of the effectiveness of the various policies at different scales.

Temporal and spatial dimensions

National scale

At the **national scale**, figures such as those published by the US Population Census Bureau (www.prb.org) require careful interpretation. Table 6.1 shows that overall in the USA the percentage of population living in urban areas has been increasing. In the UK, in the decade 1996–2006, the trend is less identifiable. But in two other advanced MEDCs, Germany and Japan, national figures also reveal an overall increase in the proportion of people living in urban areas during this period. Yet in each of these countries, especially in the USA and the UK, urban–rural migration has affected rural settlements for several decades.

Table 6.1 Percentage of the population living in urban areas in the USA, the UK, Germany and Japan, 1996–2006

Year	USA	UK	Germany	Japan
1996	75	91	85	77
1997	75	90	85	78
1998	75	90	85	78
1999	75	90	86	78
2000	75	89	86	78
2001	75	90	86	78
2002	77	90	86	78
2003	79	90	86	78
2004	79	89	88	78
2005	79	89	88	79
2006	79	89	88	79

Activity 1

(a) With reference to Table 6.1, compare the changes in percentage of urban population of the four countries shown.

(b) Explain why the figures shown in Table 6.1 can be misleading in identifying the characteristics of counterurbanisation.

(c) What would be the most effective way of representing these data graphically? Explain your choice.

Regional scale: the rural hinterlands of large urban areas

The impact of counterurbanisation and urban–rural migration has been most significant within the **rural hinterlands** of large urban areas. This is due to the high degree of social and economic interaction between urban areas and their rural hinterlands. At this scale, the effects of demographic turnaround are particularly evident. It is at this scale too that rural planning policies have been most effective.

USA

It is at the **regional scale** that counterurbanisation has been a most significant demographic trend since the 1960s. Counterurbanisation was first identified at this time in the USA, where large urban areas began to lose population while growth occurred in the immediate rural–urban fringe. Uncontrolled urban sprawl of the suburbs led to population growth beyond the defined city limits. In some areas this growth was physically contiguous with the main urban area; in others it was separated as exurban growth. Los Angeles provides a good example. Here growth occurred in Orange County and Riverside County to the east and southeast and in Ventura County to the northwest.

In the late twentieth and early twenty-first centuries, counterurbanisation has continued to have an influence in California beyond these immediate urban fringe areas. For example, rural settlements over 300 km east of San Francisco, in the foothills of the Sierra Nevada (and more recently in the high mountains), have experienced the influence of urban–rural migration. Development of tourism has been an important factor in the case of the high mountain zone.

UK

In the UK, rates of counterurbanisation and its spatial patterns have also varied considerably. In some rural hinterlands counterurbanisation was most rapid in the 1970s, with deceleration having occurred since the 1980s. This is partly explained by **reurbanisation**. In other areas, such as parts of East Anglia, counterurbanisation continues to have a significant impact on rural communities today.

The main causes of this continued rural growth have been:
- the decentralisation of employment opportunities
- improvements in transport systems, which favour commuting
- the growth in teleworking
- retirement migration

Initially, in the 1970s, this spatial pattern of growth in rural population was confined to accessible areas of the commuter hinterlands. Subsequently, growth has taken place in rural areas beyond these limits and has extended deeper into the countryside.

Thus, in some rural regions in the UK, counterurbanisation has been a relatively short-lived phenomenon. Elsewhere, urban–rural migration continues and the populations of villages and small market towns are growing as a result.

New South Wales, Australia

In New South Wales, Australia, there is clear evidence of demographic turnaround in the rural hinterlands of Sydney, Newcastle and Wollongong. The effects have been confined spatially to the coastal areas. Before 1971, there was net loss of population due to out-migration in more than 50% of the statistical local areas (SLAs) with a coastline; after 1971, there was net gain in all apart from two. Between 1971 and 1996, these coastal non-metropolitan areas increased in population by nearly 400 000 and the rural–urban fringe areas around Sydney, Newcastle and Wollongong (the largest urban centres) increased by 200 000.

Greater Taree is one example of a rural SLA on the north New South Wales coast that has experienced significant population growth despite its location in the northern extremities of the hinterlands of Newcastle and Sydney.

In the period 1991–2001, Sydney lost population from some of its inner local government areas, such as Canterbury, Marrickville and Waverley. In the same decade, Greater Taree increased in population by 6% and, since 2001, there has been a further rise of 8%. Between 1991 and 2001:

- the number of dwellings increased from 14 535 to 16 652
- the median age of residents increased from 35 to 40 years
- the percentage of professional workers increased from 11.4% to 14.5%
- the proportion employed in retailing grew from 16.9% to 18.9%
- nearly 10 000 (20%) of the workforce commuted to work by car

The main reasons for the post-1971 population growth in this rural hinterland are the attractions of environmental amenities and lifestyle in the coastal area and the growing employment opportunities (50% of the population in 2004 were of working age). Retirement has also been a significant factor; 18% of the Greater Taree population are over 65.

Activity 2

The statistics for Greater Taree suggest that urban–rural migration is not simply a trend in total population change. There have been other socioeconomic changes in this rural area to age structure, employment structure, patterns of retailing, the journey to work and house building.

Detailed information about these changes can be found online at the Australian Bureau of Statistics site www.abs.gov.au/census. The 2001 statistics for each local government area at different scales are accessible via an interactive map.

Since the latest Australian census, the statistics for 2006 are also available on the same website.

Draw up a table to compare the population and socioeconomic changes between 1991, 2001 and 2006 for:

(a) an urban area that has lost population (e.g. Marrickville, an inner local government area of Sydney)

(b) a rural area that has gained population (e.g. Greater Taree)

Village scale

At the small scale of an individual rural settlement, population turnaround is illustrated in Figure 6.1 for the village of Westonzoyland in Somerset.

Figure 6.1 Population turnaround, Westonzoyland, Somerset

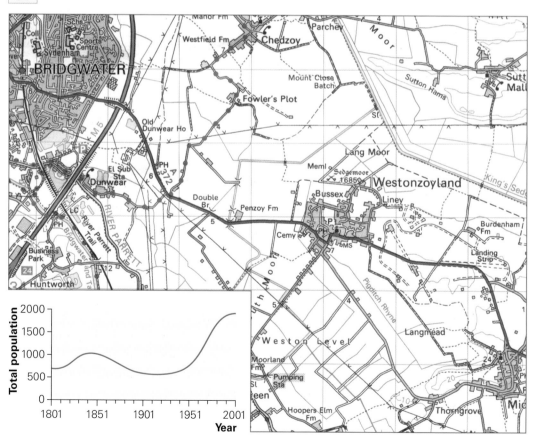

Westonzoyland is located in the rural hinterland of Bridgwater. The village experienced dramatic population growth between 1961 and 1971, following a long period of depopulation that began in the mid-nineteenth century. The simple explanation for the growth is Westonzoyland's situation: just 5 km from Bridgwater, and easily accessible along the A372, it developed rapidly as a dormitory settlement for commuters. Figure 6.1 also shows how this growth trend has decelerated since the late 1970s.

Causes of counterurbanisation

We have seen that counterurbanisation varies both spatially and temporally. These variations can be explained historically since the 1960s.

1960s: suburbanisation

Early migrations to the outer suburbs of large urban areas in the USA often caused growth in population (uncontrolled urban sprawl) beyond the defined city limits. These developments were mainly the result of dissatisfaction with the quality of life in the central urban area. This may have reflected social problems, such as increased levels of crime, degradation of the physical environment (urban blight) or health problems because of atmospheric pollution. In addition, increased car ownership and improved personal mobility allowed relocation to the rural–urban fringe for those who could afford the journey to work and the cost of suburban housing.

1970s: acceleration and exurban growth

In the 1970s, both in the UK and in the USA, we can add reasons for urban–rural migration that took place to areas physically separated from and beyond the suburbs. Access between rural and urban areas was improved with the development of the motorway network and other trunk roads. This facilitated both commuting and the movement of goods. Many individuals had residential preferences for the rural idyll. Families sought larger houses with a garden, which were initially more affordable in the countryside.

Employment opportunities were increasingly available in rural areas; this was often the result of factory owners seeking cheaper rural sites beyond the urban areas. The growth in public-sector services such as hospitals and schools in rural areas also led to increased employment opportunities.

In the 1970s, government planning agencies, such as the Location of Offices Bureau, encouraged the decentralisation of many workplaces by various fiscal and financial incentives. This attracted workforces to the countryside and was particularly effective in the London region.

Another cause of the urban–rural shift in population was slum clearance and the rehousing of populations in planned housing estates. In particular, this followed the loss of employment in manufacturing in the inner city and added further to the growth of rural population in urban fringe areas.

1980s: deceleration and twenty-first century continuation

While a slow-down of counterurbanisation has been recognised since the 1980s in some areas, the process of urban–rural migration continues to cause an increase in rural populations. The nature of rural population change depends on the stage of development of a particular region. Additional reasons currently operative in the early twenty-first century include a variety of economic and social influences.

Retirement migration is a significant factor, with particular emphasis on coastal counties in the UK (see Figure 6.2). This can be explained partly by the relatively high and increasing percentage of the UK population aged over 60. In addition, amenity migration of younger age groups is a growing trend. The main destinations are mountainous or coastal locations, where employment opportunities have developed as a result of tourism and its attendant tertiary activity. Examples include lake areas of the Sierra Nevada in California and coastal areas of New South Wales, Australia.

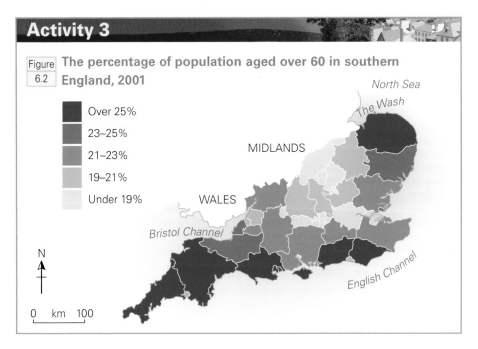

Activity 3

Figure 6.2 The percentage of population aged over 60 in southern England, 2001

Over 25%
23–25%
21–23%
19–21%
Under 19%

North Sea
The Wash
MIDLANDS
WALES
Bristol Channel
English Channel
N
0 km 100

Activity 3 (continued)

Figure 6.3   The unitary authorities of southern England

1 Plymouth
2 Torbay
3 Poole
4 Bournemouth
5 Southampton
6 Portsmouth
7 Brighton & Hove
8 North Somerset
9 Bristol
10 Bath & North East Somerset
11 South Gloucestershire
12 Swindon
13 Reading
14 Wokingham
15 Bracknell Forest
16 Windsor & Maidenhead

17 Slough
18 Thurrock
19 Medway
20 Southend
21 Buckinghamshire
22 Luton
23 Milton Keynes
24 Peterborough
25 Cambridgeshire
26 Gloucestershire
27 Bedfordshire
28 Hertfordshire
29 West Berks
30 Greater London
31 Surrey

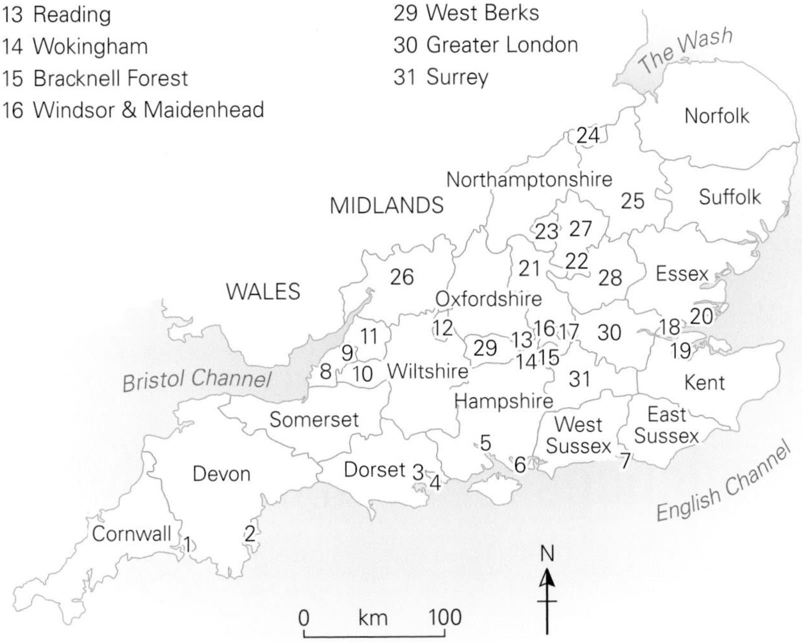

Study Figure 6.2, which shows the percentage of population aged over 60 in southern England in 2001, and Figure 6.3, the unitary authorities. With reference to Figures 6.2 and 6.3, describe the pattern of population aged over 60 in southern England. Suggest reasons for this pattern.

Teleworking from a rural retreat or even the outer suburbs is an increasingly convenient and economical practice. Many individuals make use of the internet to conduct their business, perhaps visiting their place of work much less frequently than when originally commuting. There has also been a growth in self-employed people working from their rural-based homes. The decentralisation of jobs, especially those based on modern technology, into villages and small market towns has continued to provide employment in rural areas, such as along the A14 corridor in Cambridgeshire and west Suffolk.

Falling household size is another factor. This may be the result of the increased divorce rate, which has reduced the capacity of housing stock in urban areas. There are limited sites for new housing in urban areas, except where redevelopment of brownfield sites (such as dockland areas) has encouraged reurbanisation. For this reason, demand for housing in rural areas has become increasingly desirable and necessary.

The search for the rural idyll has also continued as urban push factors have intensified. In some cities in the USA, the influx of immigrant groups into the central city areas has led to a 'white flight' to rural areas. In the Los Angeles area there are increasing developments of gated communities at some distance from Los Angeles Metropolitan County, for example in the foothills of the coastal ranges in Orange County.

Finally, it should not be forgotten that natural increase is a contributor to population growth. This is a factor particularly evident in those localities where the population structure has been rejuvenated by the influx of young families. Small market towns are a good example; they are increasingly popular, and for these younger incomers they provide not only a range of services but also good schools and relative safety in which to bring up young children.

Consequences for rural communities

Counterurbanisation in MEDCs has an impact on rural communities in many diverse ways. The effects may be categorised as demographic, economic, social and environmental. These depend on the precise location of the rural settlement within the rural hinterland and, in particular, on its degree of accessibility to the nearest large urban area. There are, for example, differences in the effects on villages in the immediate rural–urban fringe and on those in the deeper countryside.

The specific effects of counterurbanisation are best understood in the context of the original way of life, culture and rural economy of the particular area. Some effects are identified in Figure 6.4.

| Figure 6.4 | **Summary of the effects of counterurbanisation on rural communities** |

Total population growth	Suburbanisation of villages	Rise in house prices	Increasing pressure on the countryside for recreation	Increased flood potential
Change in population structure	Revival of market towns	Changes in provision of schools	Loss of shops and services in the rural–urban fringe	Enhancement of the village hall/community centre
Loss of farmland/ greenfield sites	Changes in bus services	Effects of counter-urbanisation on rural settlements	Gain of shops and services in key settlements	Activation of local councils and planning partnerships
New housebuilding	Social conflict	Development of old people's homes	Barn conversions	Wider economic base, creating employment opportunities
Growth in small-scale industry/ enterprise	Increase in traffic	Changes in provision of medical facilities	Environmental pollution	Change in employment structure

Activity 4

Study Figure 6.4.

(a) Classify each of the effects of counterurbanistion shown in Figure 6.4 as either demographic, economic, social (or socioeconomic) or environmental.

(b) Which of these effects are beneficial and which create problems for rural communities?

Rural areas have differing physical geographies and differing socioeconomic structures. Migrant flows are not always in one direction: one hinterland may experience both urban–rural and rural–urban migrations at the same time, and also include shifts in different age groups. To compound the difficulties, many rural areas have been affected by an influx of international migrants. Increasingly, the changes in population and culture brought about by counterurbanisation are causing the differences between urban and rural areas to become less distinct.

The subtlety and the structure of the planning response is therefore of vital importance to many rural communities. Most MEDCs have put in place a range of planning policies that operate at different tiers and scales. These include:

- national level; overarching bodies that fund and coordinate more local schemes
- regional and county levels
- district and local levels, which directly involve the local communities themselves

The integration of these tiers is of great significance in developing effective rural planning policies.

7 Case studies: counterurbanisation and the planning responses

This chapter includes two case studies that cover a wide variety of rural planning issues. Each case study has a structure that includes demographic change, socio-economic problems and planning/management responses, with evaluation of the outcomes.

The first case study concerns the rural hinterland of Ipswich in northeast Suffolk. This example demonstrates the complexities of planning for different types of problem within a county. The second case study deals with some recent trends in counterurbanisation in California and the associated planning issues, which to some extent are unresolved.

Northeast Suffolk

This case study outlines the problems of counterurbanisation in northeast Suffolk and assesses the effectiveness of planning responses to the problems. The study area is shown in Figure 7.1. Many planning policies have been implemented with assistance at a variety of levels, from EU to parish council. All these initiatives are effectively changing the traditional view of employment in this rural region. There is now enormous diversity and this is giving the area much greater economic stability and strength.

Demographic change

Figure 7.1 Population change in four selected villages in the rural hinterland northeast of Ipswich

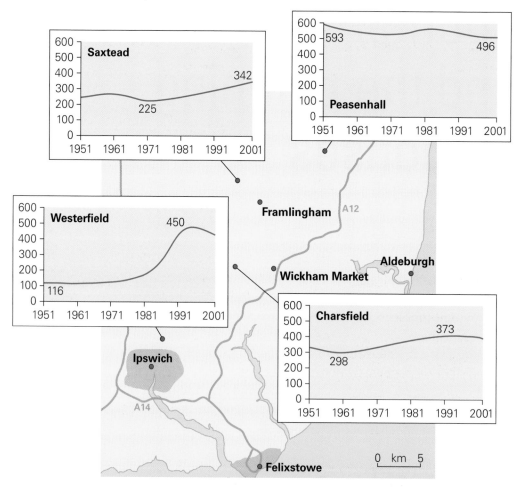

Activity 1

Study Figure 7.1, which shows population change in four villages in the rural hinterland to the northeast of Ipswich.

(a) With reference to Figure 7.1, compare and contrast the population changes of Westerfield, Charsfield, Saxtead and Peasenhall.

(b) Suggest reasons for the differences in the population changes of the villages of Westerfield and Peasenhall.

Population turnaround can be clearly identified in the rural parishes around the small market town of Framlingham.

Figure 7.2 **Population turnaround in the Framlingham area in (a) 1951–71 and (b) 1971–91**

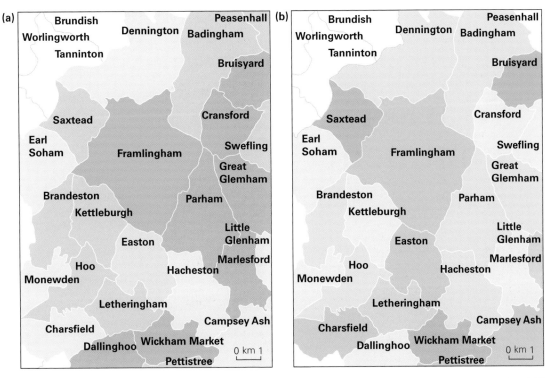

| Increase | 20+% | 10–20% | 0–10% | 0 to –10% | –10 to –20% | below –20 | Decrease |

In the 1950s and 1960s, most parishes lost population. For example, in the 1950s, Badingham lost 85 people (19%) and Bruisyard 82 (45%). This was typical of the more remote parishes, and decline continued into the 1960s. Only larger centres, such as Framlingham and Wickham Market, had a consistent increase in population.

The main cause of population change was migration. Specifically, this was due to loss of jobs in farming because of mechanisation, and to small companies moving to more accessible locations in towns. Further causes of change were:

- dissatisfaction with the rural way of life, including low wages
- few opportunities for promotion
- inadequate provision of services

A remarkable reversal of this trend occurred in the 1970s and 1980s, especially in those parishes with good access to the newly improved A12. This enabled commuting to places such as Ipswich, Martlesham, Felixstowe and the USAF air bases at Bentwaters and Woodbridge, and even London (see Figure 7.3). The underlying causes of this counterurbanisation were increased wealth and personal mobility and the better quality of life in quiet rural settlements.

| Figure 7.3 | Adapted from an article in the property section of *The Times* newspaper in the late 1990s |

Market comment

The two-hour commuting barrier kicks in as you meander through the country roads north of Ipswich. Getting to London from Peasenhall, for instance, means a 45-minute drive to Ipswich, from where the train to Liverpool Street can take 70 minutes. But the journey is manageable for weekly commuters who can work at home and limit their trips to London, and also weekenders escaping from the metropolis.

Not surprisingly, the area attracts a lot of London-based money, drawn by the appealing combination of pretty undulating countryside, the proximity of the coast and the relatively good value for money in comparison with property in the Home Counties.

Buyers want peace, seclusion, a pretty home in a pretty setting, access to sailing, golf and woodlands of the coastal area and there's a marked premium on villages that fit the bill.

Aldeburgh on the coast has its own micro values, because although it is relatively inaccessible to commuters, it is a thriving retirement centre.

Inland the gently rolling countryside of the Deben and the Alde valleys also attract outside buyers. It can be as much as 20% cheaper to buy property as you move north of Framlingham which is mainly a reflection of the distance and commuting time from London...

Negative externalities in urban areas, such as traffic congestion, pollution and higher crime rates, also played a part in this urban–rural migration (see Table 7.1).

| Table 7.1 | Population change in selected parishes of northeast Suffolk |

Parish	Population in 1971	Population in 1991	% increase 1971–1991
Bruisyard	107	180	68.2
Saxtead	225	296	31.5
Easton	292	340	16.4
Wickham Market	1436	2243	56.2
Framlingham	2258	2697	19.4

The effects of counterurbanisation

The demographic, socioeconomic and environmental consequences of counter-urbanisation can be illustrated in northeast Suffolk by the example of Framlingham. This small market town, typical of many in the UK, has experienced a dramatic transformation in the last 40 years.

Demographic change

Framlingham's population increased by 42% between 1981 and 2001 from 2190 to 3114; the estimate for the mid-census date of 2006 was 3320. The population structure has also changed significantly since 1981 (see Figure 7.4).

Figure 7.4 The population structure of Framlingham, 1981 and 2001

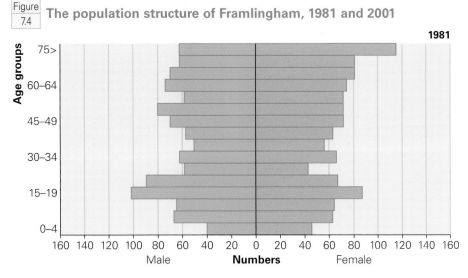

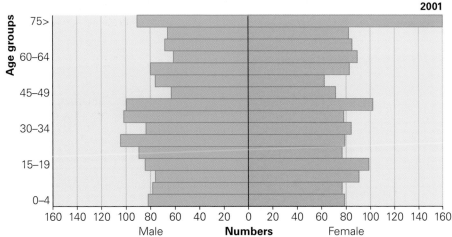

Socioeconomic changes

The socioeconomic changes include the physical expansion of new housing estates onto greenfield sites plus redevelopment of brownfield sites. Since 1980 in Framlingham there has been new residential building on 35 hectares, with 646 new houses. This includes linear development along the arterial roads (Figure 7.5(a)), infill (Figure 7.5(b)) and barn conversions.

Figure 7.5 **Typical post-1980 housing in Framlingham: (a) linear or ribbon development and (b) recent infill**

(a)

D. Barker

(b)

D. Barker

The physical expansion of Framlingham's built-up area since 1970 is entirely consistent with the three stages of development in Hudson's model of a suburbanised village (see Figure 7.6).

Figure 7.6 **Model of a suburbanised village (after Hudson)**

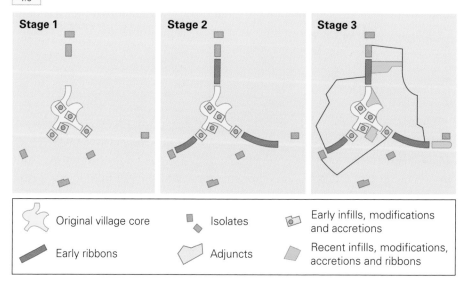

Stage 1 **Stage 2** **Stage 3**

Original village core	Isolates	Early infills, modifications and accretions
Early ribbons	Adjuncts	Recent infills, modifications, accretions and ribbons

A residential care home has been built, creating local employment for nurses, healthcare workers, cleaners and maintenance staff. The Framlingham doctors' surgery has been developed on a larger site, with five full-time doctors plus nurses. The bus service has been improved to an hourly service to Ipswich. There are four new nursery schools, two of them in converted farm premises. The town's primary and secondary schools have increased in size (classrooms and numbers on roll). There have also been various local community attempts to offer varied activities for teenagers (e.g. football club, scouts, guides, St John Ambulance and a skateboard park).

There has been an increase in the total number of shops and services, including a growth in the number of market stalls (now on a Tuesday as well as the traditional Saturday market day). Meanwhile there has been a change in the percentage of the type of shops and services from low/middle order to higher order (see Figure 7.7).

The higher economic status and demands of the new, formerly urban-based, residents help to support the seven restaurants and four hairdressers and, linked to the housing growth, four estate agents and two carpet shops. This slight cultural shift also explains the appearance of some specialist functions, such as a photographic studio, two delicatessens, a jewellers and two solicitors.

Figure 7.7 Proportions of shops and services in Framlingham, 1985 and 2005

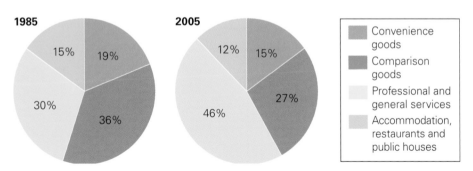

The market town's industrial and employment structures have also changed since 1980. Formerly, agriculture-based industries such as Bibby's seed merchants and Brackenbury's tarpaulin manufacturers dominated. Today there is a wider industrial base: Hatcher Components make plastic mouldings for lorries, e.g. drivers' sleeping cabins and airflow deflectors; Framlingham Technology Centre, supported by the East of England Development Agency and the EU Regional Development Fund, provides premises for start-up high-tech industries. In addition, a modern industrial estate contains 11 light industrial concerns (see Figure 7.8).

Figure 7.8

(a) Funding for Framlingham Technology Centre; (b) the River Ore light industrial estate

Framlingham has a growing numbers of teleworkers. Significantly, the number of people who commute to Ipswich and Felixstowe from Framlingham has increased.

The high demand for housing in Framlingham has inevitably led to rising house prices, often to a figure out of reach for young local residents. The median price for a three-bedroom semi-detached house was £60000 in 1990; this had risen to £180000 by 2007, at least ten times the average income of many young local people.

Environmental change

The growth of housing in Framlingham since 1980 has had a significant environmental impact. On the positive side, 65 homes have been built on former brownfield sites that were derelict agricultural-based factories. This has improved the quality of the built-up area. On the other hand, approximately 35 hectares of farmland have been lost, including nearly 2 km of hedgerows. Traffic flows have increased dramatically since 1980 as a result of commuting and the 'school run'.

Today, the River Ore, which flows through the centre of Framlingham, has higher discharge after heavy rain and the incidence of flooding has increased (see Figure 7.9). This is due to the building of new housing estates, which have increased overland flow and reduced lag times in this part of the Ore catchment.

Figure
7.9

Figure 7.9 **Bankful discharge of the River Ore prior to flooding in the suburbanised market town of Framlingham**

D. Barker

Planning/management solutions

Many of the urban–rural migrants in northeast Suffolk have high urban-based incomes and most market towns are busy and economically thriving. The coastal resorts attract income from a growing tourist trade. Figure 7.10 shows that some of the rural wards in northeast Suffolk are among the least deprived in England.

Figure 7.10 **Index of deprivation for Suffolk in 2004**

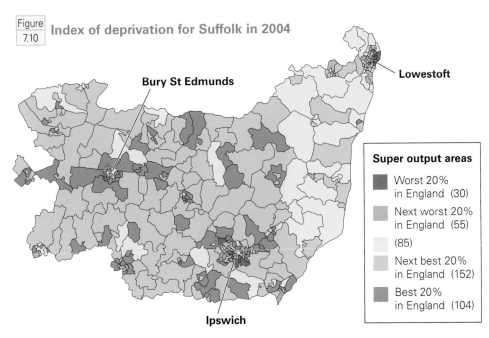

Bury St Edmunds

Lowestoft

Ipswich

Super output areas

Worst 20%
in England (30)

Next worst 20%
in England (55)

(85)

Next best 20%
in England (152)

Best 20%
in England (104)

There are, however, many localities that suffer multiple deprivation and in which resident populations are socially excluded. Some rural residents, often living within the same community as the better off, are particularly disadvantaged. These include:

- people without access to a car
- some of the elderly
- those people on lower incomes who cannot afford even to rent accommodation
- single parents
- carers of the elderly/disabled
- travellers

The statistics shown in Table 7.2 demonstrate the high level of deprivation in East Anglia. Concealed by these percentages are pockets of even more severe deprivation that threaten the sustainability of some rural settlements, such as the village of Peasenhall.

Table 7.2 Percentage of rural households at set distances from key services

	Bank <4 km	Cashpoint <4 km	Post office <2 km	Supermarket <4 km	Petrol <4 km	Primary school <2 km	Secondary school <4 km	GP <4 km	Job <8 km	Library <4 km
East Anglia	67.0	86.6	88.8	71.8	91.0	88.7	69.9	82.3	62.3	70.5
East Midlands	72.6	89.5	91.0	81.6	93.2	91.7	74.3	86.6	72.3	80.9
Northeast	78.3	88.1	90.6	82.5	90.0	89.6	77.6	84.8	75.6	79.9
Northwest	87.6	94.0	93.2	88.3	96.3	94.7	87.0	91.6	85.2	88.4
Southeast	79.0	91.5	90.8	78.0	94.5	91.8	75.4	88.7	68.0	81.5
Southwest	78.4	91.1	91.7	78.2	95.1	91.0	74.7	86.3	73.7	80.0
West Midlands	73.3	88.9	88.1	75.5	92.0	88.8	75.8	84.5	77.7	76.8
Yorkshire and Humberside	77.0	88.8	91.2	76.3	92.1	92.2	75.7	83.2	61.0	80.9

Source: Southeast Regional Research Laboratory, Birkbeck College, 2004. Commissioned analysis for the Countryside Agency.

Figure 7.11
Peasenhall: just two bus services in a week
(Newspaper article from the *East Anglian Daily Times*)

Just two bus services in a week

A village tale

PEASENHALL village may be scenic but there are few facilities for youngsters, who are forced to depend on adults with cars to take them everywhere.

The village is threatened with the closure of its small but successful primary school, which has just 14 children on its roll.

Unless it can come up with a viable scheme to federate with another primary, Suffolk County Council is set to close it at the end of this academic year.

Residents fear the closure would have a knock-on effect on other aspects of village life and make it far less attractive for young families already finding it difficult to remain in or return to their communities.

With just two bus services a week, there is little scope for older youngsters either and those who decide to seek jobs or go into further education almost inevitably have to move away from their home village if their parents don't have cars.

Mother-of-three Cheryl Baldry, of Mill Road, Peasenhall, laughed when asked about facilities for youngsters.

"There isn't any. We have not even got a playground. My daughter is 13 but there's nothing for

her. There is no transport anyway unless you have got a car to get anywhere. She'll move away eventually," she said. "Youngsters don't stay in the village because there are no jobs. They do move away. They have to learn to drive — that's essential."

Juliette Allen, of Halesworth Road, Sibton, near Peasenhall, faces similar problems with her children.

"There's nothing for the 13-year-old. You just take him everywhere. I have got an 18-year-old at Suffolk College who has to live somewhere else."

Her 18-year-old daughter can only come home at weekends because the lack of transport means that she cannot get to college if she lives at home.

Window cleaner Ed Parker, aged 30, of Bruisyard Road, Peasenhall, who has lived in the village all his life, felt public transport was a problem.

"It's absolutely non-existent, isn't it? The bus service is pathetic. They get two services once a week. Unless the parents have got transport or the young people have transport, you're just stuck."

Peasenhall parish clerk Jon James said: "I'm not too sure you can sustain living communities without the infrastructure because once the school goes, people get the idea it's not a young, living, vibrant community any more."

If it were not for government assistance and voluntary organisations, many rural settlements would be in a similar position to Peasenhall (Figure 7.11). In fact, northeast Suffolk receives assistance from an array of government bodies. Levels of funding are high and effectively sustain many rural settlements.

Several tiers of planning policy and support are available, from the scale of European Union down to the local parish council. Each tier has an essential role to play in the well-being of rural communities in northeast Suffolk.

The upper planning tiers, such as the European Regional Development Agency, DEFRA (the UK national Department for the Environment, Food and Rural Affairs) and the Countryside Agency provide financial support. They coordinate the work of lower-tier organisations and establish the basis on which rural development is directed.

On its own, however, higher-tier planning would not be entirely effective. It is important that integrated plans are developed with regional, county, district and local governments. Examples of the policies and initiatives (supported by the higher tiers of rural management) that are specifically relevant to rural settlements in northeast Suffolk include:

- Rural Priority Areas (RPAs)
- the Market Town Initiative
- promotion of tourism
- Sainsbury's Assisting Village Enterprise
- Local Strategic Partnerships
- housing for local residents
- small town and parish councils

Rural Priority Areas

Figure 7.12 shows RPAs in England. The northeast rural hinterland of Ipswich lies within the north Suffolk RPA.

| Figure 7.12 | **The RPAs of the UK** |

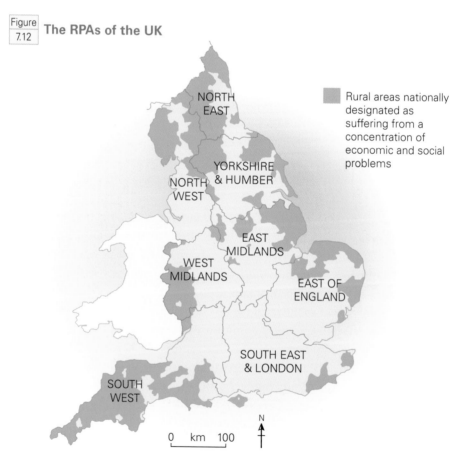

Rural areas nationally designated as suffering from a concentration of economic and social problems

0   km   100

N

Some of the RPAs qualify for assistance at the EU level and are supported under the Objective 5b programme. Under this legislation, all aspects of the rural economy receive assistance, especially where agricultural projects need finance. The northeast Suffolk area has Objective 5b status. Part of the north Suffolk RPA adjoining the Waveney Valley is particularly deprived; it has been designated, under the East Region Development Plan, as the Suffolk Objective 2 Transitional Area, which enables access to an even wider range of funding agencies (see Figure 7.13).

Figure 7.13 **The Suffolk Objective 2 Transitional Area**

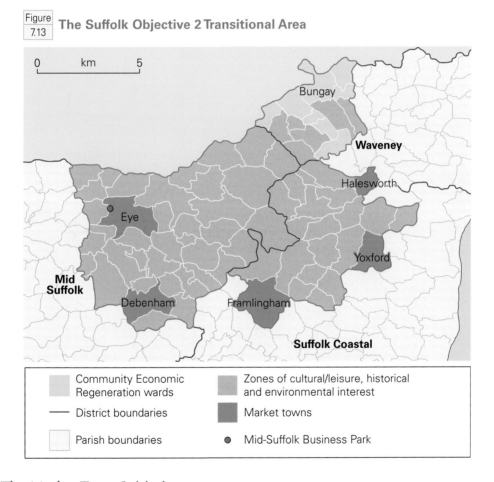

The Market Town Initiative

Framlingham is one of the small market towns in northeast Suffolk to profit from the DEFRA Market Town Initiative. It receives support since, within this RPA, it is an important service centre for an extensive surrounding hinterland (see Figure 7.14). Rural services have been concentrated in the key settlement of Framlingham to prevent total decline of basic services in the area.

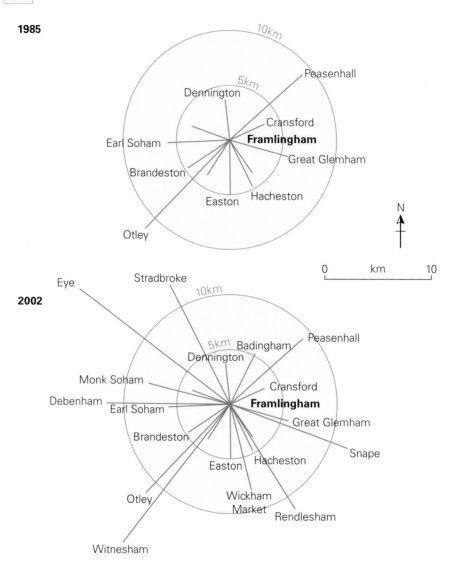

Figure 7.14 The trade areas of Framlingham, 1985 and 2002

Activity 2

Study Figure 7.14, which shows the trade areas for low- and middle-order goods and services available in Framlingham in 1985 and 2002.

(a) Compare the size and shape of Framlingham's trade areas in 1985 and 2002.

(b) Suggest three possible factors that may have influenced the change in size of the trade areas between 1985 and 2002.

Framlingham Technology Centre has been developed via the EU Objective 5b programme. It is an incubation centre (where new small-scale local businesses can set up) with accommodation for 19 firms, and the facilities are available to other small- and medium-sized enterprise high-tech companies in the area who may gain credibility from the address.

Figure 7.15 **Framlingham Technology Centre**

D. Barker

Promotion of tourism

Tourism is a growth industry. It stimulates economic activity and has an important multiplier effect. There are plans to develop the tourism industry in northeast Suffolk; the trade and income generated are expected to help sustain rural settlements.

This policy has been successful in the coastal settlements. Recent plans aim to stimulate tourism in the less popular inland area of northeast Suffolk. Public transport has been promoted by advertising the East Suffolk Line (rail link) and the interconnecting bus services, so that business is drawn to Halesworth and Yoxford, for example. The market towns of Framlingham, Debenham and Eye all have plans under Objective 5b to enhance tourism by exploiting their heritage and cultural resources. This will help to create employment, raise incomes and maintain shops and services. Establishing tourist information centres in these market towns is one way in which the tourism resources are promoted. Another is website creation, for example: www.visitsuffolk.org.uk.

Sainsbury's Assisting Village Enterprise

SAVE is a scheme designed to help the sustainability of village shops. Faced with competition from the large superstores in out-of-town locations, local shop-keepers may buy Sainsbury's products at supermarket prices. This helps to keep village shops alive. 71% of SAVE stores are found in village post offices.

Local Strategic Partnerships

Suffolk Coastal District Council has set up a Local Strategic Partnership (LSP). Its purpose is to improve the quality of life in village communities and to sustain rural settlement hierarchies.

The Suffolk Rural Services Scheme helps people who run an essential village service, such as an independent shop, pub or garage, to improve or increase the service they offer. The scheme also funds essential rural community services, such as childcare provision or the maintenance of village halls. It is funded by the East of England Development Agency, the Suffolk Development Agency, Suffolk County Council and the Suffolk Coastal LSP. Grants of up to £5000 towards meeting half the project cost are offered. Services in Peasenhall, for example, have been sustained by help received under this scheme.

Activity 3

Visit www.sclsp.org.uk or the website of a local strategic partnership operating in any rural district. Identify the role and work of the partnership in sustaining rural service provision.

Housing for local residents

Suffolk Heritage Housing Association was established in 1991. Its offices are based in Framlingham, since this is the most accessible centre for clients in housing need in northeast Suffolk. The association itself employs 120 people who live mostly within 15 km. It was established because:

- social housing is in limited supply in the area after the 'right to buy' policy
- rural employment opportunities are dwindling and incomes are low, often below the rental values of rooms/flats in the area
- second home ownership has increased
- there is a shortage of housing as numbers living alone increase (staying single longer, living longer, divorce)
- the numbers of elderly who need housing with care are increasing

The number of applicants by far exceeds the vacancy rate (only 10–12 units per year), and young people have continued to drift away from northeast Suffolk.

Small town and parish councils

Local plans are submitted to Suffolk Coastal District Council and Suffolk County Council; much depends on the initiative and vitality of the local councillors involved.

Framlingham Town Council, for example, is forward-looking and keen to ensure that this rural market town retains its high degree of centrality in northeast Suffolk, so that its own residents and those of the villages it serves remain socially included and economically supported.

Typical of many small market towns in the same situation in rural England, Framlingham promotes its flourishing market (see Figure 7.16), local shops, tourist attractions and special events designed to draw in people throughout the year. The May bank holiday gala, Midsummer Madness, the annual arts festival, an annual horse show and a Christmas charity market are just some of the events that help to sustain the settlement.

Figure 7.16 **Framlingham market**

D. Barker

California, USA

Counterurbanisation continues to be the result of urban–rural migration in many regions and states of the USA. But the early movements to the outer suburbs and the exurban areas have been superseded by significant net gains of population in areas even more distant from the cities.

To give an example, between 2000 and 2004, 18 of the 25 largest metropolitan areas in the USA experienced average annual net migration loss. The four areas with greatest loss were New York, Los Angeles, Chicago and San Francisco. In recent years, the bulk of the urban–rural flows has been into counties at least 60 km from these conurbations. This is a significant and new trend.

The highest net loss during this period was 95 000 people per year from Los Angeles County. The growing number of enclave, gated communities as far out as the foothills of the Santa Ana Mountains in the southeast of Orange County and in southwest Riverside County is another new feature of this counter-urbanisation. Moreover, the original dormitory settlements in Orange County are experiencing outflows of population, making a second migration into ever more distant rural locations.

This case study concentrates on the causes of counterurbanisation in California and its consequences for those communities more distant from the San Francisco/Oakland conurbation. These newly settled areas, often previously forest, even include parts of the Sierra Nevada, which is over 2000 m. One of the main causes of this recent trend in counterurbanisation has been changing employment opportunities. The growth of tourism, for example, in the Sierra Nevada has proved to be one of the main attractions for young migrants moving to this area.

Demographic change

The spatial pattern of urban–rural migration in California shows that increasingly distant locations have been affected by counterurbanisation since the 1960s.

At first in the 1960s, 1970s and early 1980s, the migrant flows followed the main road and rail access routes towards places such as Santa Clara to the south and Richmond to the north of the San Francisco Bay Area. These moves were driven by the increasing personal mobility afforded by a high level of car ownership, the development of freeways and also by good rail accessibility; this enabled commuting to the central areas of San Francisco.

In the late 1980s and 1990s, much of the migration away from San Francisco and other large centres, such as Oakland and Berkeley, was that of the more elderly populations. Many sought rural environments beyond the commuter belt in the foothill areas of the Sierra Nevada east of Sacramento.

In the twenty-first century, the migrations have been encouraged by the employment opportunities of tourism in the higher mountain zones. An additional factor is the large numbers of people born in the late 1940s/early 1950s baby boom, who are now relatively wealthy and seek property in retirement.

Activity 4

Figure 7.17 shows population growth in California between 1990 and 2000.

(a) Describe the pattern of population change in California.

(b) Suggest environmental, economic and social reasons for the differing rates of population growth in California.

Figure 7.17

Population change in California by county, 1990–2000

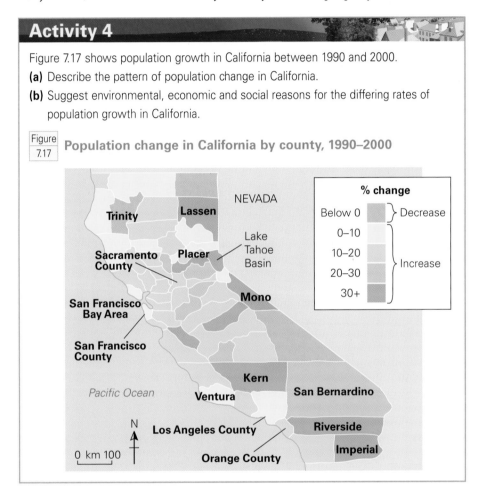

There are two specific areas affected by long distance urban–rural migration in this latter period:

- The Lake Tahoe Basin. Here, one rural settlement, aptly named Incline Village, has experienced both spatial and altitudinal expansion of permanent individual residential properties and housing estates. In 1970, the upper limit of housing was 2200 m; by 2005 it was up to 2370 m.
- The Mammoth Lakes area east of Yosemite National Park. The settlement of Mammoth Lakes has grown in population from 400 in 1960 to 7404 in 2004 and houses have now been built even above 2600 m.

| Figure 7.18 | The location of the Lake Tahoe and Mammoth Lakes areas of the central Sierra Nevada |

| Figure 7.19 | Population growth in the Lake Tahoe area, 1990–2000 |

Most of the new in-migrants are former residents of the San Francisco Bay Area and towns of the Central Valley of California. The main reasons for their migration to these rural areas are:

- the amenity value of the high Sierra; this includes the environmental quality of the mountain scenery and the opportunity for a variety of recreational activities, such as skiing, fishing, water sports and trekking
- low crime rates, with a safer environment than the city areas
- high quality of housing with less traffic, a quieter environment and greater privacy
- rising house prices in the Bay Area, pushed up by reurbanisation demand from young single people and 'empty-nesters'; this induces many families to take the opportunity to gain capital by selling their urban home and moving to an area with relatively low housing costs in the mountains
- new forms of telecommunication, encouraging decentralisation of some urban businesses
- 'ethnic homogeneity' (80% of the residents are white) — a major reason cited by many incomers
- more easily obtained planning permission to build; there are vast, wide-open spaces of private land available for development, which have not been protected by the Federal or State government as national forests, national parks, wilderness or ecological areas
- employment opportunities in tourism, such as skiing and gaming in the casino hotels, including those across the California/Nevada border

Socioeconomic problems

There are many consequences of this urban–rural migration for the rural settlements of the high Sierra Nevada, including the need for a variety of planning responses.

There have been changes in age–sex structure. Whereas the foothills of the Sierra Nevada are popular with the retired, the high Sierra is dominated by the younger age groups. For example, in Truckee (north of Lake Tahoe) 42% of the population are between the ages of 20 and 44, and there is a significant male domination (183:100 in the 20–34 age group).

Employment structure has changed since 1970. By 2004, in the Lake Tahoe region, the percentage of the working population employed in agriculture and forestry had fallen to just 1%. Jobs in the tertiary sector linked to tourism predominate in the districts close to Lake Tahoe and, further south in Mammoth Lakes, almost 35% of the working population have at least a first degree. What future difficulties are these changes in population and employment structures likely to cause?

The most dramatic effect of counterurbanisation is that of rising house prices. There has been a rapid increase in prices in California as a whole, but this has been exceeded easily in both the Lake Tahoe and Mammoth Lakes areas, as shown in Figure 7.20.

Figure 7.20 Median house prices in California and the high central Sierra Nevada, 2004

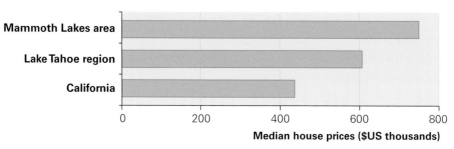

Median house prices ($US thousands)

There has been conflict between the in-migrants, who tend to be high-income groups, and the low-income local population, who are often forced out to the nearby towns where housing prices are lower. Housing prices are being driven up in the popular lakeside wooded settlement of Incline Village, where the median price for a single family residence in 2006 was $1.08 million. In 2007, prices for houses with three bedrooms ranged from $0.75 million to $1.4 million, while prices for detached houses with four bedrooms were between $2 million and $3 million. Provision of affordable housing for local people is the most serious issue of socioeconomic sustainability in both Incline Village and Mammoth Lakes.

The growth in number and range of both low- and high-order services has increased the centrality of Incline Village. This settlement has, apart from its main grocery store and other convenience shops, a community hospital, high, middle and elementary schools, a courthouse and a library, plus sports shops, dentists, and a hardware and furniture store. These all depend on high-threshold populations, which are sustained by the resident population of 9952 (2004) and by more than double this figure in the number of annual visitors.

Planning/management solutions

Two problems in this mountain environment are managed efficiently; a third remains a difficulty.

Seasonal unemployment is often a problem in mountain areas but this is being dealt with in Incline Village. Year-round alternative tourist activities have been organised by the local community, including yachting, golf, horseback riding, hiking (e.g. via the Tahoe Rim Trail Association) and fitness activities in the Incline

Village Recreation Centre. All of these offer a wide range of local employment opportunities. This is a common type of development in smaller settlements and is a most effective response at the lowest tier of rural management.

The pollution of Lake Tahoe is a potential problem as building and tourism grow. The lake is situated in two states (California and Nevada). Formerly, the separate county plans were less effective in managing the natural ecosystems of the lake basin. But today the bi-state Tahoe Regional Planning Agency, established in 1969, coordinates planning for the entire physical lake system. In addition, the Lake Tahoe Environmental Education Coalition monitors fish stocks and other aquatic life, and assists in the control of eutrophication and levels of acidity.

A remaining planning problem for both Incline Village and Mammoth Lakes is that of physical sprawl of the built-up areas into mountain habitats. Residents of both settlements seek low-density accommodation either as permanent or holiday homes. In 2002, 38% of houses around Lake Tahoe and 44% in Mammoth Lakes were second homes.

The physical spread of low-density house construction runs contrary to the successful 'smart growth' planning policies of the San Francisco Bay Area. These smart gowth policies effectively limit suburbanisation using green belt restrictions on urban sprawl, encourage reurbanisation by revitalising city centres, charge developers the full costs of environmental harm, or provide better loan terms for new buildings near public transport. Significantly, conservation programmes are established to prevent valued land from being developed.

Local plans such as the Mammoth Lakes Community Plan do include reference to smart growth and the need for environmental stewardship, but there is still pressure on the landscapes and ecosystems of the central Sierra as a result of this relatively uncontrolled growth. As yet there has been little guidance from planners at the higher levels to restrict or encourage the necessary concentration of housing development in this area.

Activity 5

The following questions on rural settlements require essay-style answers.

1 (a) Describe how retirement migration and the increase in the number of second and holiday homes have affected rural services.

   (b) Explain the measures taken in different regions to address the issue of changing demand for rural services.

2 To what extent do second homes have a negative influence on rural areas in MEDCs?

Activity 5 (continued)

3 **(a)** Describe the main changes in the location of public services, such as schools and healthcare, in rural areas over the past 40 years.

   (b) Explain how the interaction of several factors brings about changes in rural service provision.

4 **(a)** What is meant by 'hierarchy' and 'centrality' in the location of service activities?

   (b) Explain the changes in many service hierarchies during the past 40 years.

5 How important is a sustainable hierarchy of rural service centres to the economic and social well-being of rural communities in MEDCs?

6 **(a)** Describe how rural service provision varies from one area to another.

   (b) Explain how planning responses have addressed issues caused by rural service decline.

7 'Rural population changes in MEDCs in the past 30 years have generated important economic and social issues.' With reference to named rural areas, describe these major issues and discuss the effectiveness of planning responses to them.

8 Why are many rural communities in MEDCs currently in decline? How successful have governments been in halting this decline?